William Robert Grove

The Correlation of Physical Forces

William Robert Grove

The Correlation of Physical Forces

ISBN/EAN: 9783337775759

Printed in Europe, USA, Canada, Australia, Japan

Cover: Foto ©berggeist007 / pixelio.de

More available books at **www.hansebooks.com**

THE CORRELATION

OF

PHYSICAL FORCES.

BY

W. R. GROVE, Q.C., M.A., F.R.S.

CORRESPONDING MEMBER OF THE ACADEMIES OF
ROME, TURIN, ETC.

FIFTH EDITION.

FOLLOWED BY A DISCOURSE ON

CONTINUITY

DELIVERED BY THE AUTHOR AS
PRESIDENT OF THE BRITISH ASSOCIATION AT NOTTINGHAM, MDCCCLXVI.

LONDON:
LONGMANS, GREEN, AND CO.
1867.

PREFACE.

THE phrase 'Correlation of Physical Forces' in the sense in which I have used it, having become recognised by a large number of scientific writers, it would produce confusion were I now to adopt another title. It would, perhaps, have been better if I had in the first instance used the term Co-relation, as the words 'correlate,' 'correlative,' had acquired a peculiar metaphysical sense somewhat differing from that which I attached to the substantive correlation. The passage in the text (p. 247) explains the meaning I have given to the term.

Twenty-five years having elapsed since I promulgated the views contained in this Essay, which were first advanced in a lecture at the London Institution in January 1842, printed by the proprietors, and subsequently more fully developed in a course of lectures in 1843, published in abstract in the 'Literary Gazette,' I think it advisable to add a little to the Preface with reference to other labourers in the same field.

It has happened with this subject as with many others, that similar ideas have independently presented themselves to different minds about the same period. In 'Leibig and Wohler's Annalen' for May 1842,* a paper appeared by M. Mayer which I had not read when my third edition was published, but which I have now read in the translation by Mr. Youmans of New York. It deduces very much the same conclusions to which I had been led, the author starting partly from à priori reasoning and partly from an experiment by which water was heated by agitation, and from another, which had, however, previously been made by Davy, viz. that ice can be melted by friction, though kept in a medium which is below the freezing point of water.

In 1843 a paper by Mr. Joule on the mechanical equivalent of heat appeared, which, though not in terms touching on the mutual and necessary dependence of all the Physical Forces, yet bears most importantly upon the doctrine.

While my third edition was going through the press I had the good fortune to make the acquaint-

* I am informed that these papers are, in fact, published some time after the date at which they appear in the magazine.

ance of M. Seguin, who informed me that his uncle, the eminent Montgolfier, had long entertained the idea that force was indestructible, though, with the exception of one sentence in his paper on the hydraulic ram, and where he is apparently speaking of mechanical force, he has left nothing in print on the subject. Not so, however, M. Seguin himself, who in 1839, in a work on the 'Influence of Railroads,' has distinctly expressed his uncle's and his own views on the identity of heat and mechanical force, and has given a calculation of their equivalent relation, which is not far from the more recent numerical results of Mayer, Joule, and others.

Several of the great mathematicians of a much earlier period advocated the idea of what they termed the Conservation of Force, but although they considered that a body in motion would so continue for ever, unless arrested by the impact of another body, and, indeed, in the latter case, would, if elastic, still continue to move (though deflected from its course) with a force proportionate to its elasticity, yet with inelastic bodies the general and, as far as I am aware, the universal belief was, that the motion was arrested and the force annihilated. Montgolfier went a step farther, and his hydraulic ram was to him a proof of the truth of his preconceived idea, that the shock or impact of bodies left the mechanical force undestroyed.*

Previously, however, to the discoveries of the voltaic battery, electro-magnetism, thermo-electricity, and photography, it was impossible for any mind to perceive what, in the greater number of cases, became of the force which was apparently lost. The phenomena of heat, known from the earliest times, would have been a mode of accounting for the resulting force in many cases where motion was arrested, and we find Bacon enouncing a theory that motion was the form, as he quaintly termed it, of heat. Rumford and Davy adopted this view, the former with a fair approximative attempt at numerical calculation, but no one of these philosophers seems to have connected it with the indestructibility of force. A passage in the writings of Dr. Roget, combating the theory that mere contact of dissimilar bodies was the source of voltaic electricity, philosophically supports his argument by the idea of non-creation of force.

As I have introduced into the later editions of my Essay abstracts of the different discoveries which I have found, since my first lectures, to bear upon

---

* See also a paper by Mr. Rankine, 'On the Dynamical Principles of Newton,' in the 'Engineer,' Oct. 26, 1866.

the subject, I have been regarded by many rather as the historian of the progress made in this branch of thought than as one who has had anything to do with its initiation. Everyone is but a poor judge where he is himself interested, and I therefore write with diffidence, but it would be affecting an indifference which I do not feel if I did not state that I believe myself to have been the first who introduced this subject as a generalised system of philosophy, and continued to enforce it in my lectures and writings for many years, during which it met with the opposition usual and proper to novel ideas.

Avocations necessary to the well-being of others have prevented my following it up experimentally, to the extent that I once hoped; but I trust and believe that this Essay, imperfect though it be, has helped materially to impress on that portion of the public which devotes its attention rather to the philosophy of science than to what is now termed science, the truth of the thesis advocated.

To show that the work of to-day is not substantially different from the thoughts I first published on the subject, at a period when I knew little or nothing of what had been thought before, I venture to give a few extracts from the printed copy of my lecture of 1842:—

PREFACE.

Physical Science treats of Matter, and what I shall to-night term its *Affections*; namely, Attraction, Motion, Heat, Light, Electricity, Magnetism, Chemical-Affinity. When these re-act upon matter, they constitute Forces. The present tendency of theory seems to lead to the opinion that all these Affections are resolvable into one, namely, Motion: however, should the theories on these subjects be ultimately so effectually generalised as to become laws, they cannot avoid the necessity for retaining different names for these different Affections; or, as they would then be called, different Modes of Motion.
. . . .

Œrsted proved that Electricity and Magnetism are two forces which act upon each other; not in straight lines, as all other known forces do, but in a rectangular direction: that is, that bodies invested with electricity, or the conduits of an electric-current, tend to place magnets at right angles to them; and, conversely, that magnets tend to place bodies conducting electricity at right angles to them. . . . .

The discovery of Œrsted, by which electricity was made a source of Magnetism, soon led philosophers to seek the converse effect; that is, to educe Electricity from a permanent magnet:— had these experimentalists succeeded in their expectations of making a stationary magnet a source of electric-currents, they would have realised the ancient dreams of perpetual motion, they would have converted statics into dynamics, they would have produced power without expenditure; in other words, they would have become creators. They failed, and Faraday saw their error : he proved that to obtain Electricity from Magnetism it was necessary to superadd to this latter, motion ; that magnets while in motion induced electricity in contiguous conductors; and that the direction of such electric-currents was tangential to the polar direction of the magnet ; that as Dynamic-electricity may be made the source of Magnetism and Motion, so Magnetism conjoined with Motion may be made the source of Electricity. Here originates the Science of Magneto-electricity, the true converse of Electro-magnetism; and thus between Electricity and Magnetism is shewn to exist a reciprocity of force such that, considering either as the primary agent, the other becomes the re-agent; viewing one in the relation of cause, the other is the effect. . . . .

The Science of Thermo-Electricity connected heat with electricity, and proved these, like all other natural forces, to be capable of mutual reaction. . . . .

Voltaic action is Chemical action taking place at a distance, or transferred through a chain of media: and the Daltonian equivalent numbers are the exponents of the amount of voltaic action for corresponding chemical substances. . . . .

By regarding the quantity of electrical, as directly proportional to the efficient chemical action, and by experimentally tracing this principle, I have been fortunate enough to increase the power of the Voltaic-pile more than sixteen times, as compared with any combination previously known. . . . .

I am strongly disposed to consider that the facts of Catalysis depend upon voltaic action, to generate which three heterogeneous substances are always necessary: Induced by this belief I made some experiments on the subject, and succeeded in forming a voltaic combination by gaseous-oxygen, gaseous-hydrogen, and platinum; by which a galvanometer was deflected and water decomposed. . . . .

It appears to me that heat and light may be considered as affections; or, according to the Undulatory-theory, vibrations of matter itself, and not of a distinct ethereal fluid permeating it: these vibrations would be propagated, just as sound is propagated by vibrations of wood or as waves by water. To my mind, all the consequences of the Undulatory-theory flow as easily from this, as from the hypothesis of a specific ether; to suppose which, namely, to suppose a fluid *sui generis*, and of extreme tenuity penetrating solid bodies, we must assume, first, the existence of the fluid itself; secondly, that bodies are without exception porous; thirdly, that these pores communicate; fourthly, that matter is limited in expansibility. None of these difficulties apply to the modification of this theory which I venture to propose; and no other difficulty applies to it which does not equally apply to the received hypothesis. With regard to the planetary spaces, the diminishing periods of comets is a strong argument for the existence of an universally-diffused matter: this has the function of resistance, and there appears to be no reason to divest it of the functions common to all matter, or specifically to appropriate it to certain affections. Again, the phenomena of transparency and opacity are, to my mind, more easily explicable by the former than by the latter theory; as resulting from a difference in the molecular arrangement of the matter affected. In regard to the effects of double-refraction and polarisation, the molecular structure gives at once a reason for the effects upon the one theory, while upon the other we must, in addition to previous assumptions, further assume a different elasticity of the ether in different directions within the doubly-refracting medium. The same theory is applicable to Electricity and Magnetism; my own experiments on the influence of the elastic intermedium on the voltaic-arc, and those of Faraday on electrical induction, furnish strong arguments in support of it. My inclination would lead me to detain you on this subject much longer than my judgment deems advisable: I therefore content myself with offering it to your consideration, and should my avocations permit, I may at a future period more fully develope it. . . . .

Light, Heat, Electricity, Magnetism, Motion, and Chemical-affinity, are all convertible material affections; assuming either as the cause, one of the others will be the effect: thus heat may be said to produce electricity, electricity to produce heat; magnetism to produce electricity, electricity magnetism; and so of the rest. Cause and effect, therefore, in their abstract relation to these forces, are words solely of convenience: we are totally unacquainted with the ultimate generating power of each and all of them, and probably shall ever remain so; we can only ascertain the normæ of their action: we must humbly refer their causation to one omnipresent influence, and content ourselves with studying their effects and developing by experiment their mutual relations.

I have transposed the passages relating to voltaic action and catalysis, but I have not added a word to the above quotations, and, as far as I am now aware, the theory that the so-called imponderables are affections of ordinary matter, that they are resolvable into motion, that they are to be regarded in their action on matter as forces, and not as specific entities, and that they are capable of mutual reaction, thence alternately acting as cause and effect, had not at that time been publicly advanced.*

My original Essay being a record of lectures, and being published by the managers of the Institution, I necessarily adhered to the form and matter which I had orally communicated. In preparing subsequent editions I found that, without destroying the identity of the work, I could not alter the style; although it would have been less difficult and more satisfactory to me to have done so, the work would not then have been a republication; and I was for

* See also an experiment shown at the London Institution, p. 297 post.

obvious reasons anxious to preserve as far as I could the original text, which, though added to, is but little altered.

The form of lectures has necessarily continued the use of the first person, and I would beg my readers not to attribute to me, from the modes of expression used, a dogmatism which is far from my thought. If my opinions are expressed broadly, the reason is that when opinions are always hedged in by qualifications, the style becomes embarrassed and the meaning frequently unintelligible.

As the main object of a course of lectures is to induce the auditor to think, and to consult works on the subject he hears treated, so the object of this Essay is more to induce a particular train of thought on the known facts of physical science than to enter with minute criticism into each separate branch.

In one or two of the reviews of previous editions the general idea of the work was objected to. I believe, however, that will not now be the case; the mathematical labours of Mr. Thompson, Clausius, and others, though not suitable for insertion in an Essay such as this, have awakened an interest for many portions of the subject, which promises much for its future progress.

The short and irregular intervals which my profession permits me to devote to science so prevent the continuity of attention necessary for the proper evolution of a train of thought, that I certainly should not now have courage to publish for the first time such an Essay; and it is only the favour it has received from those whose opinions I highly value, and the, I trust pardonable, wish not to let some favourite thoughts of my youth lose all connection with my name, that have induced me to reprint it.

My scientific readers will, I hope, excuse the very short notices of certain branches of science which are introduced, as without them the work would be unintelligible to many for whom it is intended. I have endeavoured so to arrange the subjects that each division should form an introduction to those which follow, and to assume no more preliminary knowledge to be possessed by my readers than would be expected from persons acquainted with the elements of physical science.

The notes contain references to the original memoirs in which the branches of science alluded to are to be found, as well as to those which bear on the main arguments; where these memoirs are numerous, or not easy of access, I have referred to treatises in which they are collated. To prevent the reader's attention being interrupted, I have in the notes referred to the pages of the text, instead of to interpolated letters.

For this fifth edition I have revised the whole of the text, and several correspondents having suggested that I should add to it my address as President of the British Association, 1866, it seemed to me on consideration not inappropriate. In it are noticed several new discoveries relating to the subjects treated of in this Essay, which, had I not decided on adding the discourse, I should have incorporated with the text. The doctrine of continuity, however, though closely allied with that of correlation, could not well have been interwoven with the Essay, and is more suitable as a sequel.

I cannot but feel gratified with the reception that address has met with, and although the portions referring to the continuity of succession of organised beings have, as I rather expected, been the subject of some adverse comment, yet as I have seen no argument against them, I have nothing at present to answer or to alter.

Several of the arguments adduced by me were written some years ago, and before the appearance of Darwin's own celebrated work. I had then no notion of the effects of natural selection in modifying organisms, but in order to test as fairly as I could the reasons for and against continuity, as opposed to special creations, I wrote down at different times in the form of a dialogue everything that occurred to me as bearing most strongly on each side of the question, and showed it to several friends with whom I discussed the subject. My resulting opinion is given in the text.

I thought I had sufficiently guarded myself, in the passage at p. 334, from being supposed to deny that there are or have been catastrophes or cataclysms; but it appears from some comments that I have not done so. If sea or river undermine a cliff and the cliff fall, it is undoubtedly a cataclysm; if I tread on a beetle, it is a catastrophe to the beetle; but formation and destruction are very different things, and the tenour of my discourse applies to genesis not extinction. Even in phenomena such as those of Geology, though there were doubtless cataclysms, and sometimes on a much larger scale than at other times, yet the evidence seems to me to point to their being limited in extent at any one period, when compared with the whole terrestrial surface.

CORRELATION

OF

PHYSICAL FORCES.

WHEN natural phenomena are for the first time observed, a tendency immediately developes itself to refer them to something previously known—to bring them within the range of acknowledged sequences. The mode of regarding new facts, which is most favourably received by the public, is that which refers them to recognised views—stamps them into the mould in which the mind has been already shaped. The new fact may be far removed from those to which it is referred, and may belong to a different order of analogies, but this cannot then be known, as its co-ordinates are wanting. It may be questionable whether the mind is not so moulded by past events that it is impossible to advance an entirely new view, but admitting such possibility, the new view, necessarily founded on insufficient data, is likely to be more incorrect and prejudicial than even a strained attempt to reconcile the new discovery with known facts.

The theory consequent upon new facts, whether it be a co-ordination of them with known ones, or the more difficult and dangerous attempt at re-modelling the public ideas, is generally enunciated by the discoverers themselves of the facts, or by those to whose authority the world at the period of the discovery defers; others are not bold enough, or if they be so, are unheeded. The earliest theories thus enunciated obtain the firmest hold upon the public mind, for at such a time there is no power of testing, by a sufficient range of experience, the truth of the theory; it is accepted solely or mainly upon authority: there being no means of contradiction, its reception is, in the first instance, attended with some degree of doubt, but as the time in which it can fairly be investigated far exceeds that of any lives then in being, and as neither the individual nor the public mind will long tolerate a state of abeyance, a theory shortly becomes, for want of a better, admitted as an established truth: it is handed from father to son, and gradually takes its place in education. Succeeding generations, whose minds are thus formed to an established view, are much less likely to abandon it. They have adopted it in the first instance, upon authority, to them unquestionable, and subsequently to yield up their faith would involve a laborious remodelling of ideas, a task which the public as a body will and can rarely undertake, the frequent occurrence of which is indeed inconsistent with the very existence of man in a social state, as it would induce an anarchy of thought—a perpetuity of mental revolutions.

This necessity has its good ; but the prejudicial effect upon the advance of science is, that by this means, theories the most immature frequently become the most permanent ; for no theory can be more immature, none is likely to be so incorrect, as that which is formed at the first flush of a new discovery ; and though time exalts the authority of those from whom it emanated, time can never give to the illustrious dead the means of analysing and correcting erroneous views which subsequent discoveries confer.

Take for instance the Ptolemaic System, which we may almost literally explain by the expression of Shakspeare : ' He that is giddy thinks the world turns round.' We now see the error of this system, because we have all an immediate opportunity of refuting it ; but this identical error was received as a truth for centuries, because, when first promulgated, the means of refuting it were not at hand, and when the means of its refutation became attainable, mankind had been so educated to the supposed truth, that they rejected the proof of its fallacy.

I have premised the above for two reasons: first, to obtain a fair hearing, by requesting as far as possible a dismissal from the minds of my readers of preconceived views by and in favour of which all are liable to be prejudiced; and secondly, to defend myself from the charge of undervaluing authority, or treating lightly the opinions of those to whom and to whose memory mankind looks with reverence. Properly to value authority we should estimate it together with its means of information: if 'a dwarf on the shoulders of a giant can see further than the giant,' he is no less a dwarf in comparison with the giant.

The subject on which I am about to treat—viz., the relation of the affections of matter to each other and to matter—peculiarly demands an unprejudiced regard. The different aspects under which these agencies have been contemplated; the different views which have been taken of matter itself; the metaphysical subtleties to which these views unavoidably lead, if pursued beyond fair inductions from existing experience, present difficulties almost insurmountable.

The extent of claim which my views on this subject may have to originality have been stated in the Preface; they became strongly impressed upon my mind at a period when I was much engaged in experimental research, and were, as I then believed, and still believe, regarding them as a system, new: expressions in the works of different authors, bearing more or less on the subject, have subsequently been pointed out to me, some of which go back to a distant period. An attempt to analyse these in detail, and to trace how far I have been anticipated by others, would probably but little interest the reader, and in the course of it I should constantly have to make distinctions showing wherein I differed, and wherein I agreed with others. I might cite authorities which appear to me to oppose, and others which appear to coincide with certain of the views I have put forth; but this would interrupt the consecutive developement of my own ideas, and might render me liable to the charge of misconstruing those of others; I therefore think it better to avoid such discussion in the text; and in addition to the sketch given in the Preface, to furnish in the notes at the conclusion such references to different authors as bear upon the subjects treated of, which I have discovered, or which have been pointed out to me since the delivery of the lectures of which this essay is a record.

The more extended our research becomes, the more we find that knowledge is a thing of slow progression, that the very notions which appear to ourselves new, have arisen, though perhaps in a very indirect manner, from successive modifications of traditional opinions. Each word we utter, each thought we think, has in it the vestiges, is in itself the impress, of antecedent words and thoughts. As each material form, could we rightly read it, is a book, containing in itself the past history of the world; so, different though our philosophy may now appear to be from that of our progenitors, it is but theirs added to or subtracted from, transmitted drop by drop through the filter of antecedent, as ours will be through that of subsequent, ages.—The relic is to the past as is the germ to the future.

Though many valuable facts, and correct deductions from them, are to be found scattered amongst the voluminous works of the ancient philosophers; yet, giving them the credit which they pre-eminently deserve for having devoted their lives to purely intellectual pursuits, and for having thought, seldom frivolously, often profoundly, nothing can be more difficult than to seize and apprehend the ideas of those who reasoned from abstraction to abstraction—who, although, as we now believe, they must have depended upon observation for their first inductions, afterwards raised upon them such a complex superstructure of syllogistic deductions, that, without following the same paths, and tracing the same sinuosities which led them to their conclusions, such conclusions are to us unintelligible. To think as another thought, we must be placed in the same situation as he was placed: the errors of commentators generally arise from their reasoning upon the arguments of their text, either in blind obedience to its dicta, without considering the circumstances under which they were uttered, or in viewing the images presented to the original writer from a different point to that from which he viewed them. Experimental philosophy keeps in check the errors both of *à priori* reasoning and of commentators, and, at all events, prevents their becoming cumulative; though the theories or explanations of a fact be different, the fact remains the same. It is, moreover, itself the exponent of its discoverer's thought: the observation of known phenomena have led him to elicit from nature the new phenomenon: and though he may be wrong in his deductions from this after its discovery, the reasonings which conducted him to it are themselves valuable, and, having led from known to unknown truths, can seldom be uninstructive.

Very different views existed amongst the ancients as to the aims to be pursued by physical investigation, and as to the objects likely to be attained by it. I do not here mean the moral objects, such as the attainment of the *summum bonum*, &c. —but the acquisitions in knowledge which such investigations were likely to confer. Utility was one object in view, and this was to some extent attained by the progress made in astronomy and mechanics; Archimedes, for instance, seems to have constantly had this end in view; but, while pursuing natural knowledge for the sake of knowledge, and the power which it brings with it, the greater number seemed to entertain an expectation of arriving at some ultimate goal, some point of knowledge, which would give them a mastery over the mysteries of nature, and would enable them to ascertain what was the most intimate structure of matter, and the causes of the changes it exhibits. Where they could not discover, they speculated. Leucippus, Democritus, and others, have given us their notions of the ultimate atoms of which matter was formed, and of the *modus agendi* of nature in the various transformations which matter undergoes.

The expectation of arriving at ultimate causes or essences continued long after the speculations of the ancients had been abandoned, and continues even to the present day to be a very general notion of the objects to be ultimately attained by physical science. Francis Bacon, the great remodeller of science, entertained this notion, and thought that, by experimentally testing natural phenomena, we should be enabled to trace them to certain primary essences or causes whence the various phenomena flow. These he speaks of under the scholastic name of 'forms'—a term derived from the ancient philosophy, but differently applied. He appears to have understood by 'form' the essence of quality — that in which, abstracting everything extraneous, a given quality consists, or that which, superinduced on any body, would give it its peculiar quality : thus the form of transparency is that which constitutes transparency, or that by which, when discovered, transparency could be produced or superinduced. To take a specific example of what I may term the synthetic application of his philosophy : — ' In gold there meet together yellowness, gravity, malleability, fixedness in the fire, a determinate way of solution, which are the simple natures in gold; for he who understands form, and the manner of superinducing this yellowness, gravity, ductility, fixedness, faculty of fusion, solution, &c., with their particular degrees and proportions, will consider how to join them together in some body, so that a transmutation into gold shall follow.

On the other hand, the analytic method, or, ' the enquiry from what origin gold or any other metal or stone is generated from its first fluid matter or rudiments, up to a perfect mineral,' is to be perceived by what Bacon calls the latent process, or a search for 'what in every generation or transformation of bodies, flies off, what remains behind, what is added, what separated, &c.; also, in other alterations and motions, what gives motion, what governs it, and the like.' Bacon appears to have thought that qualities separate from the substances themselves were attainable, and if not capable of physical isolation, were at all events capable of physical transference and superinduction.

Subsequently to Bacon a belief has generally existed, and now to a great extent exists, in what are called secondary causes, or consequential steps, wherein one phenomenon is supposed necessarily to hang on another, and that on another, until at last we arrive at an essential cause, subject immediately to the First Cause. This notion is generally prevalent both on the Continent and in this country: nothing is more familiar than the expression 'study the effects in order to arrive at the causes.'

Instead of regarding the proper object of physical science as a search after essential causes, I believe it ought to be, and must be, a search after facts and relations — that although the word Cause may be used in a secondary and concrete sense, as meaning antecedent forces, yet in an abstract sense it is totally inapplicable: we cannot predicate of any physical agency that it is abstractedly the cause of another; and if, for the sake of convenience, the language of secondary causation be permissible, it should be only with reference to the special phenomena referred to, as it can never be generalised.

The misuse, or rather varied use, of the term

Cause, has been a source of great confusion in physical theories, and philosophers are even now by no means agreed as to their conception of causation. The most generally received view of causation, that of Hume, refers it to invariable antecedence — i. e., we call that a cause which invariably precedes, that an effect which invariably succeeds. Many instances of invariable sequence might however be selected, which do not present the relation of cause and effect: thus, as Reid observes and Brown does not satisfactorily answer, day invariably precedes night, and yet day is not the cause of night. The seed, again, precedes the plant, but is not the cause of it; so that when we study physical phenomena it becomes difficult to separate the idea of causation from that of force, and these have been regarded as identical by some philosophers. To take an example which will contrast these two views: if a floodgate be raised, the water flows out; in ordinary parlance, the water is said to flow *because* the floodgate is raised: the sequence is invariable; no floodgate, properly so called, can be raised without the water flowing out, and yet in another, and perhaps more strict, sense, it is the gravitation of the water which *causes* it to flow. But though we may truly say that, in this instance, gravitation causes the water to flow, we cannot in truth abstract the proposition, and say, generally, that gravitation is the cause of water flowing, as water may flow from other causes, gaseous elasticity, for instance, which will cause water to flow from a receiver full of air into one that is exhausted; gravitation may also, under certain circumstances, arrest instead of cause the flow of water.

Upon neither view, however, can we get at anything like abstract causation. If we regard causation as invariable sequence, we can find no case in which a given antecedent is the only antecedent to a given sequent: thus, if water could flow from no other cause than the withdrawal of a floodgate, we might say abstractedly that this was the cause of water flowing. If, again, adopting the view which looks to causation as a force, we could say that water could be caused to flow only by gravitation, we might say abstractedly that gravitation was the cause of water flowing — but this we cannot say; and if we seek and examine any other example, we shall find that causation is only predicable of it in the particular case, and cannot be supported as an abstract proposition; yet this is constantly attempted. Nevertheless, in each *particular* case where we speak of Cause, we habitually refer to some antecedent power or force: we never see motion or any change in matter take effect without regarding it as produced by some previous change; and, when we cannot trace it to its antecedent, we mentally refer it to one; but whether this habit be philosophically correct is by no means clear. In other words, it seems questionable, not only whether cause and effect are convertible terms with antecedence and sequence, but whether in fact cause does precede effect, whether force does precede the change in matter of which it is said to be the cause.

The actual priority of cause to effect has been doubted, and their simultaneity argued with much ability. As an instance of this argument it may be said, the attraction which causes iron to approach the magnet is simultaneous with and ever accompanies the movement of the iron; the movement is evidence of the coexisting cause or force, but there is no evidence of any interval in time between the one and the other. On this view time would cease to be a necessary element in causation; the idea of cause, except perhaps as referred to a primeval creation, would cease to exist; and the same arguments which apply to the simultaneity of cause with effect would apply to the simultaneity of Force with Motion. We could not, however, even if we adopted this view, dispense with the element of time in the sequence of phenomena; the effect being thus regarded as ever accompanied simultaneously by its appropriate cause, we should still refer it to some antecedent effect; and our reasoning as applied to the successive production of all natural changes would be the same.

Habit and the identification of thoughts with phenomena so compel the use of recognised terms, that we cannot avoid using the word cause even in the sense in which objection is taken; and if we struck it out of our vocabulary, our language, in speaking of successive changes, would be unintelligible to the present generation. The common error, if I am right in supposing it to be such, consists in the abstraction of cause, and in supposing in each case a general secondary cause—a something which is not the first cause, but which, if we examine it carefully, must have all the attributes of a first cause, and an existence independent of, and dominant over, matter.

The relations of electricity and magnetism afford us a very instructive example of the belief in secondary causation. Subsequent to the discovery by Oersted of electro-magnetism, and prior to that by Faraday of magneto- electricity, electricity and magnetism were believed by the highest authorities to stand in the relation of cause and effect— i.e. electricity was regarded as the cause, and magnetism as the effect; and where magnets existed without any apparent electrical currents to cause their magnetism, hypothetical currents were supposed, for the purpose of carrying out the causative view; but magnetism may now be said with equal truth to be the cause of electricity, and electrical currents may be referred to hypothetical magnetic lines: if therefore electricity cause magnetism, and magnetism cause electricity, why then electricity causes electricity, which becomes, so as to speak, a *reductio ad absurdum* of the doctrine.

To take another instance, which may render these positions more intelligible. By heating bars of bismuth and antimony in contact, a current of electricity is produced; and if their extremities be united by a fine wire, the wire is heated. Now here the electricity in the metals is said to be caused by heat, and the heat in the wire to be caused by electricity, and in a concrete sense this is true; but can we thence say abstractedly that heat is the cause of electricity, or that electricity is the cause of heat? Certainly not; for if either be true, both must be so, and the effect then becomes the cause of the cause, or, in other words, a thing causes itself. Any other proposition on this subject will be found to involve similar difficulties, until, at length, the mind will become convinced that abstract secondary causation does not exist, and that a search after essential causes is vain.

The position which I seek to establish in this Essay is, that the various affections of matter which constitute the main objects of experimental physics, viz., heat, light, electricity, magnetism, chemical affinity, and motion, are all correlative, or have a reciprocal dependence: that neither, taken abstractedly, can be said to be the essential cause of the others, but that either may produce or be convertible into, any of the others: thus heat may mediately or immediately produce electricity, electricity may produce heat; and so of the rest, each merging itself as the force it produces becomes developed: and that the same must hold good of other forces, it being an irresistible inference from observed phenomena that a force cannot originate otherwise than by devolution from some pre-existing force or forces.

The term force, although used in very different senses by different authors, in its limited sense may be defined as that which produces or resists motion. Although strongly inclined to believe that the other affections of matter, which I have above named, are and will ultimately be resolved into, modes of motion, many arguments for which view will be given in subsequent parts of this Essay, it would be going too far, at present, to assume their identity with it; I therefore use the term force in reference to them, as meaning that active principle inseparable from matter which is supposed to induce its various changes.

The word force and the idea it aims at expressing might be objected to by the purely physical philosopher on similar grounds to those which apply to the word cause, as it represents a subtle mental conception, and not a sensuous perception or phenomenon. The objection would take something of this form. If the string of a bent bow be cut, the bow will straighten itself; we thence say there is an elastic *force* in the bow which straightens it; but if we applied our expressions to this experiment alone, the use of the term force would be superfluous, and would not add to our knowledge on the subject. All the information which our minds could get would be as sufficiently obtained from the expression, when the string is cut the bow becomes straight, as from the expression, the bow becomes straight by its elastic force. Do we know more of the phenomenon, viewed without reference to other phenomena, by saying it is produced by force? Certainly not. All we know or see is the effect; we do not see force—we see motion or moving matter.

If now we take a piece of caoutchouc and stretch it, when released it returns to its original length. Here, though the subject-matter is very different, we see some analogy in the effect or phenomenon to that of the strung bow. If again we suspend an apple by a string, cut the string, the apple falls. Here, though it is less striking, there is still an analogy to the strung bow and the caoutchouc.

Now when the word force is employed as comprehending these three different phenomena we find some use in the term, not by its explaining or rendering more intelligible the *modus agendi* of matter, but as conveying to the mind something which is alike in the three phenomena, however distinct they may be in other respects: the word becomes an abstract or generalised expression, and regarded in this light is of high utility. Although I have given only three examples, it is obvious that the term would equally apply to 300 or 3,000 cases.

But it will be said, the term force is used not as expressing the effect, but as that which produces the effect. This is true, and in this its ordinary sense I shall use it in these pages. But though the term has thus a potential meaning, to depart from which would render language unintelligible, we must guard against supposing that we know essentially more of the phenomena by saying they are produced by something, which something is only a word derived from the constancy and similarity of the phenomena we seek to explain by it. The relations of the phenomena to which the terms force or forces are applied give us real knowledge; these relations may be called relations of forces; our knowledge of them is not thereby lessened, and the convenience of expression is greatly increased, but the separate phenomena are not more intimately known; no further insight into why the apple falls is acquired by saying it is forced to fall, or it falls by the force of gravitation; by the latter expression we are enabled to relate it most usefully to other phenomena, but we still know no more of the particular phenomenon than that under certain circumstances the apple does fall.

In the above illustrations, force has been treated as the producer of motion, in which case the evidence of the force is the motion produced; thus we estimate the force used to project a cannon ball in terms of the mass of matter, and the velocity with which it is projected. The evidence of force when the term is applied to resistance to motion is of a somewhat different character; the matter resisting is molecularly affected, and has its structure more or less changed; thus a strip of caoutchouc to which a weight is suspended is elongated, and its molecules are displaced as compared with their position when unaffected by the gravitating force. So a piece of glass bent by an appended weight has its whole structure changed; this internal change is made evident by transmitting through it a beam of polarised light; a relation thus becomes established between the molecular state of bodies and the external forces or motion of masses. Every particle of the caoutchouc or glass must be acting and contributing to resist or arrest the motion of the mass of matter appended to it.

It is difficult, in such cases, not to recognise a reality in force. We need some word to express this state of tension; we know that it produces an effect, though the effect be negative in character: although in this effort of inanimate matter we can no more trace the mode of action to its ultimate elements than we can follow out the connection of our own muscles with the volition which calls them into action, we are experimentally convinced that matter changes its state by the agency of other matter, and this agency we call force.

In placing the weight on the glass, we have moved the former to an extent equivalent to that which it would again describe if the resistance were removed, and this motion of the mass becomes an exponent or measure of the force exerted on the glass; while this is in the state of tension, the force is ever existing, capable of reproducing the original motion, and while in a state of abeyance as to actual motion, it is really acting on the glass. The motion is suspended, but the force is not annihilated.

But it may be objected, if tension or static force be thus motion in abeyance, there is at all times a large amount of dynamical action subtracted from the universe. Every stone raised and left upon a hill, every spring that is bent, and has required force to upraise or bend it, has for a time, and possibly for ever, withdrawn this force, and annihilated it. Not so; when we raise a weight and leave it at the point to which it has been elevated we have changed the centre of gravity of the earth, and consequently the earth's position with reference to the sun, planets, and stars; the effort we have made pervades and shakes the universe; nor can we present to the mind any exercise of force, which is not thus permanent in its dynamical effects. If, instead of one weight being raised, we raise two weights, each placed at points of the earth diametrically opposite each other, it would be said, here you have compensation, a balance, no change in the centre of gravity of the earth; but we have increased the mean diameter of the earth, and a perturbation of our planet, and of all other celestial bodies, necessarily ensues.

The force may be said to be in abeyance with reference to the effect it would have produced, if not arrested, or placed in a state of tension; but in the act of imposing this state, the relations of equilibrium with other bodies have been changed, and these move in their turn, so that motion of the same amount would seem to be ever affecting matter conceived in its totality.

Press the hands violently together; the first notion may be that this is power locked up, and that no change ensues. Not so; the blood courses more quickly, respiration is accelerated, changes, which we may not be able to trace, take place in the muscles and nerves, transpiration is increased; we have given off force in various ways, and must, if the effort be prolonged, replenish our sources of power, by fresh chemical action in the stomach.

In books which treat of statics and dynamics, it is common and perhaps necessary to isolate the subjects of consideration; to suppose, for instance, two bodies gravitating, and to ignore the rest of the universe. But no such isolation exists in reality, nor could we predict the result if it did exist. Would two bodies gravitate towards each other in empty space, if space can be empty? the notion that they would is founded on the theory of attraction, which Newton himself repudiated, further than as a convenient means of regarding the subject. For purposes of instruction or argument it may be convenient to assume isolated matter: many conclusions so arrived at may be true, but many will be erroneous.

If, in producing effects of tension or of static force, the effort made pervades the universe, it may be said, when the bent spring is freed, when the raised weight falls, a converse series of motions must be effected, and this theory would lead to a mere reciprocation, which would be equally unproductive of permanent change with the annihilation of force. If raising the weight has changed the centre of gravity of the earth, and thence of the universe, the fall of the weight, it will be said, restores the original centre of gravity, and everything comes back to its original status. In this argument we again, in thought, isolate our experiment; we neglect surrounding circumstances. Between the time of the raising and falling of the weight, be the interval never so small, nay, more, during the rising and during the fall, the earth has been going on revolving round its axis and round the sun, to say nothing of other changes, such as temperature, cosmical magnetism, &c., which we may call accidental, but which, if we knew all, would probably be found to be as necessary and as reducible to law as the motion of the earth. A change having taken place, the fall of the weight does not bring back the *status quo*, but other changes supervene, and so on. Nothing repeats itself, because nothing can be placed again in the same condition: the past is irrevocable.

MOTION.

Motion—which has been taken as the main exponent of force in the above examples—is the most obvious, the most distinctly conceived of all the affections of matter. Visible motion, or relative change of position in space, is a phenomenon so obvious to simple apprehension, that to attempt to define it would be to render it more obscure; but with motion, as with all physical appearances, there are certain vanishing gradations or undefined limits, at which the obvious mode of action fades away; to detect the continuing existence of the phenomena we are obliged to have recourse to other than ordinary methods of investigation, and we frequently apply other and different names to the effects so recognised.

Thus sound is motion; and although in the earlier periods of philosophy the identity of sound and motion was not traced out, and they were considered distinct affections of matter—indeed, at the close of the last century a theory was advanced that sound was transmitted by the vibrations of an ether—we now so readily resolve sound into motion, that to those who are familiar with acoustics, the phenomena of sound immediately present to the mind the idea of motion, i.e. motion of ordinary matter.

Again, with regard to light: no doubt now exists that light moves or is accompanied by motion. Here the phenomena of motion are not made evident by the ordinary sensuous perception, as for instance the motion of a visibly moving projectile would be, but by an inverse deduction from known relations of motion to time and space: as all observation teaches us that bodies in moving from one point in space to another occupy time, we conclude that, wherever a continuing phenomenon is rendered evident in two different points of space at different times, there is motion, though we cannot see the progression. A similar deduction convinces us of the motion of electricity.

As we in common parlance speak of sound moving, although sound is motion, it requires no great stretch of imagination to conceive light and electricity as motions, and not as things moving. If one end of a long bar of metal be struck, a sound is soon perceptible at the other end. This we now know to be a vibration of the bar; sound is but a word expressive of the mode of motion impressed on the bar; so one end of a column of air or glass subjected to a luminous impulse gives a perceptible effect of light at the other end: this can equally be conceived to be a vibration or transmitted motion of particles in the transparent column: this question will, however, be further discussed hereafter; for the present we will confine ourselves to motion within the limits to which the term is usually restricted.

With the perceptible phenomena of motion the mental conception has been invariably associated to which I have before alluded, and to which the term force is given—the which conception, when we analyse it, refers us to some antecedent motion. If we except the production of motion by heat, light, &c., which will be considered in the sequel, when we see a body moving we look to motion having been communicated to it by matter which has previously moved.

Of absolute rest Nature gives us no evidence: all matter, as far as we can ascertain, is ever in movement, not merely in masses, as with the planetary spheres, but also molecularly, or throughout its most intimate structure: thus every alteration of temperature produces a molecular change throughout the whole substance heated or cooled; slow chemical or electrical actions, actions of light or invisible radiant forces, are always at play, so that as a fact we cannot predicate of any portion of matter that it is absolutely at rest. Supposing, however, that motion is not an indispensable function of matter, but that matter can be at rest, matter at rest would never of itself cease to be at rest; it would not move unless impelled to such motion by some other moving body, or body which has moved. This proposition applies not merely to impulsive motion, as when a ball at rest is struck by a moving body, or pressed by a spring which has previously been moved, but to motion caused by attractions such as magnetism or gravitation. Suppose a piece of iron at rest in contact with a magnet at rest; if it be desired to move the iron by the attraction of the magnet, the magnet or the iron must first be moved; so before a body falls it must first be raised. A body at rest would therefore continue so for ever, and a body once in motion would continue so for ever, in the same direction and with the same velocity, unless impeded by some other body, or affected by some other force than that which originally impelled it. These propositions may seem somewhat arbitrary, and it has been doubted whether they are necessary truths; they have for a long time been received as axioms, and there can at all events be no harm in accepting them as postulates. It is however very generally believed that if the visible or palpable motion of one body be arrested by its impact on another body, the motion ceases, and the force which produced it is annihilated.

Now the view which I venture to submit is, that force cannot be annihilated, but is merely subdivided or altered in direction or character. First, as to direction. Wave your hand: the motion, which has apparently ceased, is taken up by the air, from the air by the walls of the room, &c., and so by direct and reacting waves, continually comminuted, but never destroyed. It is true that, at a certain point, we lose all means of detecting the motion, from its minute subdivision, which defies our most delicate means of appreciation, but we can indefinitely extend our power of detecting it accordingly as we confine its direction, or increase the delicacy of our examination. Thus, if the hand be moved in unconfined air, the motion of the air would not be sensible to a person at a few feet distance; but if a piston of the same extent of surface as the hand be moved with the same rapidity in a tube, the blast of air may be distinctly felt at several yards distance. There is no greater absolute amount of motion in the air in the second than in the first case, but its direction is restrained, so as to make the means of detection more facile. By carrying on this restraint, as in the air-gun, we get a power of detecting the motion, and of moving other bodies at far greater distances. The puff of air which would in the air-gun project a bullet a quarter of a mile, if allowed to escape without its direction being restrained, as by the bursting of a bladder, would not be perceptible at a yard distance, though the same absolute amount of motion be impressed on the surrounding air.

It may, however, be asked, what becomes of force when motion is arrested or impeded by the counter-motion of another body? This is generally believed to produce rest, or entire destruction of motion, and consequent annihilation of force: so indeed it may, as regards the motion of the masses, but a new force, or new character of force, now ensues, the exponent of which, instead of visible motion, is heat. I venture to regard the heat which results from friction or percussion as a continuation of the force which was previously associated with the moving body, and which, when this impinges on another body, ceasing to exist as gross, palpable motion, continues to exist as heat.

Thus, let two bodies, A and B, be supposed to move in opposite directions (putting for the moment out of the question all resistance, such as that of the air, &c.), if they pass each other without contact each will move on for ever in its respective direction with the original velocity, but if they touch each other the velocity of the movement of each is reduced, and each becomes heated: if this contact be slight, or such as to occasion but a slight diminution of their velocity, as when the surfaces of the bodies are oiled, then the heat is slight; but if the contact be such as to occasion a great diminution of motion, as in percussion, or as when the surfaces are roughened, then the heat is great, so that in all cases the resulting heat is proportionate to the diminished velocity. Where, instead of resisting and consequently impeding the motion of the body A, the body B gives way, or itself takes up the motion originally communicated to A, then we have less heat in proportion to the motion of the body B, for here the operation of the force continues in the form of palpable motion: thus the heat resulting from friction in the axle of a wheel is lessened by surrounding it by rollers; these take up the primary motion of the axle, and the less, by this means, the initial motion is impeded, the less is the resulting heat. Again, if a body move in a fluid, although some heat is produced, the heat is apparently trifling, because the particles of the fluid themselves move, and continue the motion originally communicated to the moving body: for every portion of motion communicated to them this loses an equivalent, and where both lose, then an equivalent of heat results.

As the converse of this proposition, it should follow that the more rigid the bodies impinging on each other the greater should be the amount of heat developed by friction, and so we find it. Flint, steel, hard stones, glass, and metals, are those bodies which give the greatest amount of heat from friction or percussion; while water, oil, &c., give little or no heat, and from the ready mobility of their particles lessen its developement when interposed between rigid moving bodies. Thus, if we oil the axles of wheels, we have more rapid motion of the bodies themselves, but less heat; if we increase the resistance to motion, as by roughening the points of contact, so that each particle strikes against and impedes the motion of others, then we have diminished motion, but increased heat; or if the bodies be smooth, but instead of sliding past each other be pressed closely together and then rubbed, we shall in many cases evolve more heat than by the roughened bodies, as we get a greater number of particles in contact and a greater resistance to the initial motion. I cannot present to my mind any case of heat resulting from friction which is not explicable by this view: friction, according to it, is simply impeded motion. The greater the impediment, the more force is required to overcome it, and the greater is the resulting heat; this resulting heat being a continuation of indestructible force, capable, as we shall presently see, of reproducing palpable motion, or motion of definite masses.

Whatever be the nature of the bodies, rough or smooth, solid or liquid, provided there be the same initial force, and the whole motion be ultimately arrested, there should be the same amount of heat developed, though where the motion is carried on through a great number of points of matter we do not so sensibly perceive the resulting heat from its greater dissipation. The friction of fluids produces heat, an effect first noticed I believe by Mayer. The total heat produced by the friction of fluids should, therefore, it will be said, be equal to that produced by the friction of solids; for although each particle produces little heat, the motion being readily taken up by the neighbouring particles, yet by the time the whole mass has attained a state of rest there has been the same impeding of the initial motion as by the friction of solids if produced by the same initial force. If the heat be viewed in the aggregate, and allowance be made for the specific thermal capacity of the substances employed, it probably is the same, though apparently less; the heat in the case of solids being manifested at certain defined points, while in that of fluids it is dissipated, both the time and space during and through which the motion is propagated differ in the two cases, so that the heat in the latter case is more readily carried off by surrounding bodies.

If the body be elastic, and by its reaction the motion impressed on it by the initial force be continued, then the heat is proportionately less; and were a substance perfectly elastic, and no resistance opposed to it by the air or other matter, then the movement once impressed would be perpetual, and no heat would result. A ball of caoutchouc bandied about for many minutes between a racket and a wall is not perceptibly heated, while a leaden bullet projected by a gun against a wall is rendered so hot as to be intolerable to the touch : in the former case, the motion of the mass is continued by the reaction due to its elasticity; in the latter the motion of the mass is extinguished and heat ensues.

A pendulum started in the exhausted receiver of an air-pump continues its oscillation for hours or even days ; the friction at its point of suspension and the resistance of the air is minimised, and the heat is therefore imperceptible, but these trifling resistances in the end arrest the motion of the mass, the one giving it out as heat, the other conveying the force to the receiver, and thence to surrounding bodies. Similar reasoning may be applied to the oscillation of a coiled spring and balance wheel.

To wind up a clock a certain amount of force is expended by the arm; this force is given back by the descent of the weight, the wheels move, the pendulum is kept oscillating, heat is generated at each point of friction, and the surrounding air is set in motion, a part of which is made obvious to us by the ticking sound. But it will be said, if instead of allowing the weight to act upon the machinery, the cord by which it is suspended be cut, the weight drops and the force is at an end. By no means, for in this case the house is shaken by the concussion, and thus the force and motion are continued, while in the former case the weight reaches the ground quietly, and no evidence of force or motion is manifested by its impact, the whole having been previously dissipated.

If the initial motion, instead of being arrested by the impact of other bodies, as in friction or percussion, is impeded by confinement or compression, as where the dilatation of a gas is prevented by mechanical means, heat equally results: thus if a piston is used to compress air in a closed vessel, the compressed air and, from it, the sides of the vessel will be heated: the air being unable to take up and carry on the original motion communicates molecular motion or expansion to all bodies in contact with it; and, conversely, if we expand air by mechanical motion, as by withdrawing the piston, cold is produced. So when a solid has its particles compressed or brought nearer together, as when a bar of iron is hammered, heat is produced beyond that which is due to percussion alone. In this latter case we cannot very easily effect the converse result, or produce cold by the mechanical dilatation of a solid, though the phenomena of solution, where the particles of a solid are detached from each other, or drawn more widely asunder, give us an approximation to it: in the case of solution cold is produced.

We are from a very extensive range of observation and experiment entitled to conclude that, with some curious exceptions to be presently noticed, whenever a body is compressed or brought into smaller dimensions it is heated, i.e. it expands neighbouring substances. Whenever it is dilated or increased in volume it is cooled, or contracts neighbouring substances.

Mr. Joule has made a great number of experiments for the purpose of ascertaining what quantity of heat is produced by a given mechanical action. His mode of experimenting is as follows. An apparatus formed of floats or paddles of brass or iron is made to rotate in a bath of water or mercury. The power which gives rise to this rotation is a weight raised like a clock-weight to a certain height; this by acting during its fall on a spindle and pulley communicates motion to the paddle-wheel, the water or mercury serving as a friction medium and calorimeter; and the heat is measured by a delicate mercurial thermometer. The results of his experiments he considers prove that a fall of 772 lbs. through a space of one foot is able to raise the temperature of one pound of water through one degree of Fahrenheit's thermometer. Mr. Joule's experiments are of extreme delicacy—he tabulates to the thousandth part of a degree of Fahrenheit, and a large number of his thermometric data are comprehended within the limits of a single degree. Other experimenters have given very different numerical results, but the general opinion seems to be that the numbers given by

Mr. Joule are the nearest approximation to the truth yet obtained.

Hitherto I have taken no distinction as to the physical character of the bodies impinging on each other; but Nature gives us a remarkable difference in the character or mode of the force eliminated by friction, accordingly as the bodies which impinge are homogeneous or heterogeneous: if the former, heat alone is produced; if the latter, electricity.

We find, indeed, instances given by authors, of electricity resulting from the friction of homogeneous bodies; but, as I stated in my original Lectures, I have not found such facts confirmed by my own experiments, and this conclusion has been corroborated by some experiments of Professor Erman, communicated to the meeting of the British Association in the year 1845, in which he found that no electricity resulted from the friction of perfectly homogeneous substances; as, for instance, the ends of a broken bar. Such experiments as these will, indeed, be seldom free from slight electrical currents, on account of the practical difficulty of fulfilling the condition of perfect homogeneity in the substances themselves, their size, their temperature, &c. ; but the effects produced are very trifling and vary in direction, and the resultant effect is nought. Indeed, it would be difficult to conceive the contrary. How could we possibly image to the mind or describe the direction of a current from the same body to the same body, or give instructions for a repetition of the experiment? It would be unintelligible to say that in rubbing to and fro two pieces of bismuth, iron, or glass, a current of electricity circulated from bismuth to bismuth, or from iron to iron, or from glass to glass; for the question immediately occurs —from which bismuth to which does it circulate? And should this question be answered by calling one piece A, and the other B, this would only apply to the particular specimens employed, the distinctive appellation denoting a distinction in fact, as otherwise A could be substituted for B, and the bar to which the positive electricity flowed would in turn become the bar to which the negative electricity flowed. We may say that it circulates from rough glass to smooth, from cast iron to wrought, for here there is not homogeneity. It is moreover conceivable, that when the motion is continuous in a definite direction, electricity may result from the friction of homogeneous bodies. If A and B rub against each other, revolving in opposite directions, concentric currents of positive and negative electricity may be conceived circulating within the metals, and be described by reference to the direction of their motion; this indeed would be a different phenomenon from those we have been considering; but without some distinction between the two substances in quality or direction, the electrical effects are indescribable, if not inconceivable.

When, however, homogeneous bodies are fractured or even rubbed together, phenomena are observed to which the term electricity is applied; a flash or line of light appears at the point of friction which by some is called electrical, by others phosphorescent.

I have myself observed a remarkable case of the kind in the caoutchouc fabric now commonly used for waterproof clothing: if two folds of this substance be allowed to cohere so as partly to unite and present a difficulty of separation, then, on stripping the one from the other, or tearing them asunder, a line of light will follow the line of separation.

If this class of phenomena be electrical, it is electricity determined as it is generated; there is no dual character impressed on the matter acting, the flash is electrical as a spark from the percussion of flint is electrical, or as the slow combustion of phosphorus, or any other case of the developement of heat and light. It seems to be better to class this phenomenon under the categories of heat and light than under that of electricity, the latter word being retained for those cases where a dual or polar character of force is manifested. In experiments which have been made by the friction of similar substances where the one appears positively and the other negatively electrical, there will be found some difference in the mode of rubbing by which the molecular state of the bodies is in all probability changed, making one a dissimilar substance from the other; thus it is said by Bergmann, that when two pieces of glass are rubbed so that all the parts of one pass over one part of the other, the former is positive and the latter negative. It is obvious that in this case the rubbing in one is confined to a line, and that must be more altered in molecular structure at the line of friction than the one where the friction is spread over the whole surface: so if a ribbon be drawn transversely over another ribbon, the substances are not, *qua* the rubbing action, identical; so again, in the rupture of crystals, we are dealing with substances having a polar arrangement of particles — the surfaces of the fragments cannot be assumed to be molecularly identical.

The developement of electricity by the common electrical machine arises, as far as I can understand it, from the separation or rupture of contiguity between dissimilar bodies; a metallic surface, the amalgam of the cushion, is in contact with glass; these two bodies act upon each other by the force of cohesion; and when, by an external mechanical force, this is ruptured, as it is at each moment of the motion of the glass plate or cylinder, electricity is developed in each: were they similar bodies, heat only would be developed.

According to the experiments of Mr. Sullivan electricity may be produced by vibration alone if the substance vibrating be composed either of dissimilar metals, as a wire partly of iron and partly of brass, caused to emit a musical sound, or of the same metal if its parts be not homogeneous, as a piece of iron one portion of which is hard and crystallised and the other soft and fibrous; the current resulting appears to be due to the vibration, and not to heat engendered, as it ceases immediately with the vibration.

We may say, then, that in our present state of knowledge, where the mutually impinging bodies are homogeneous, heat and not electricity is the result of friction and percussion; where the bodies impinging are heterogeneous, we may safely state that electricity is always produced by friction or percussion, although heat in a greater or less degree accompanies it; but when we come to the question of the ratio in which frictional electricity is produced, as determined by the different characters of the substances employed, we find very complex results. Bodies may differ in so many particulars which influence more or less the developement of electricity, such as their chemical constitution, the state of their surfaces, their state of aggregation, their transparency or opacity, their power of conducting electricity, &c., that the *normæ* of their action are very difficult of attainment. As a general rule, it may be said that the developement of electricity is greater when the substances employed are broadly distinct in their physical and chemical qualities, and more particularly in their conducting powers; but up to the present time the laws governing such developement have not been even approximately determined.

I have said, in reference to the various forces or affections of matter, that either of them may, *mediately* or *immediately*, produce the others ; and this is all I can venture to predicate of them in the present state of science ; but after much consideration I incline strongly to the opinion that science is rapidly progressing towards the establishment of immediate or direct relations between all these forces. Where at present no immediate relation is established between any of them, electricity generally forms the intervening link or middle term.

Motion, then, will directly produce *heat* and *electricity*, and electricity, being produced by it, will produce *magnetism*—a force which is always developed by electrical currents at right angles to the direction of those currents, as will be subsequently more fully explained. *Light* also is readily produced by motion, either directly, as when accompanying the heat of friction, or mediately, by electricity resulting from motion; as in the electoral spark, which has most of the attributes of solar light, differing from it only in those respects in which light differs when emanating from different sources or seen through different media; for instance, in the position of the fixed lines in the spectrum or in the ratios of the spaces occupied by rays of different refrangibility. In the decompositions and compositions which the terminal points proceeding from the conductors of an electrical machine develope when immersed in different chemical media, we get the production of *chemical affinity* by electricity, of which motion is the initial source. Lastly, *motion* may be again reproduced by the forces which have emanated from motion; thus, the divergence of the electrometer, the revolution of the electrical wheel, the deflection of the magnetic needle, are, when resulting from frictional electricity, palpable movements reproduced by the intermediate modes of force, which have themselves been originated by motion.

HEAT.

IF we now take HEAT as our starting point, we shall find that the other modes of force may be readily produced by it. To take motion first: this is so generally, I think I may say invariably, the immediate effect of heat, that we may almost, if not entirely, resolve heat into motion, and view it as a mechanically repulsive force, a force antagonistic to attraction of cohesion or aggregation, and tending to *move* the particles of all bodies, or to separate them from each other.

It may be well here to premise, that in using the terms 'particles,' or 'molecules,' which will be frequently employed in this Essay, I do not use them in the sense of the atomist, or mean to assert that matter consists of indivisible particles or atoms. The words will be used for the necessary purpose of contradistinguishing the action of the indefinitely minute physical elements of matter from that of masses having a sensible magnitude, much in the same way as the terms 'lines' or 'points' may be used, and with advantage in an abstract sense; though there does not exist, in fact, a thing which has length and breadth without thickness, and though a thing without parts or dimensions is nothing.

If we put aside the sensation which heat produces in our own bodies, and regard heat simply in relation to its effect upon inorganic matter, we find that, with a very few exceptions, which I shall presently notice, the effects of what is called heat are simply an expansion of the matter acted upon, and that the matter so expanded has the power by its own contraction of communicating expansion to all bodies in contiguity with it. Thus, if the body be a solid: for instance, iron; a liquid, say water; or a gas, say atmospheric air—each of these, when heated, is expanded in every direction: in the two former cases, by increasing the heat to a certain point, we change the physical character of the substance, the solid becomes a liquid, and the liquid becomes a gas; these, however, are still expansions, particularly the latter, when, at a certain period, the expansion becomes rapidly and indefinitely greater. But what is, in fact, commonly done in order to heat a substance, or to increase the heat of a substance? it is merely approximated to some other heated, that is, to some other expanded substance, which latter is cooled or contracted as the former expands. Let us now divest the mind of the impression that heat is in itself anything substantive, and suppose that these phenomena are regarded for the first time, and without any preconceived notions on the subject; let us introduce no hypothesis, but merely express as simply as we can the facts of which we have become cognisant; to what do they amount ? to this, that matter has pertaining to it a molecular repulsive power, a power of dilatation, which is communicable by contiguity or proximity.

Heat thus viewed, is motion, and this molecular motion we may readily change into the motion of masses, or motion in its most ordinary and palpable form : for example, in the steam engine, the piston and all its concomitant masses of matter are moved by the molecular dilatation of the vapour of water.

To produce continuous motion there must be an alternate action of heat and cold ; a given portion of air, for instance, heated beyond the temperature of the circumambient air, is expanded. If now it be made to act on a movable piston, it moves this to a point at which the tension or elastic force of the confined air equals that of the surrounding air. If the confined air be kept at this point, the piston would remain stationary; but if it be cooled, the external air exercising then a greater relative degree of pressure, the piston returns towards its original position; just as it will be seen, when we come to the magnetic force, that a magnet placed in a particular position produces motion in iron near it, but to make this motion continuous, or to obtain an available mechanical power, the magnet must be demagnetised, or else a stable equilibrium is obtained.

In the case of the piston moved by heated air the motion of the mass becomes the exponent of the amount of heat—i. e. of the expansion or separation of the molecules; and we do not, by any of our ordinary methods, test heat in any other way than by its purely dynamical action. The various modifications of the thermometer and pyrometer are all measurers of heat by motion: in these instruments liquid or solid bodies are expanded and elongated, i. e. moved in a definite direction, and, either by their own visible motion, or by the motion of an attached index, communicate to our senses the amount of the force by which they are moved. There are, indeed, some delicate experiments which tend to prove that a repulsive action between separate masses is produced by heat. Fresnel found that mobile bodies heated in an exhausted receiver repelled each other to sensible distances; and Baden Powell found that the coloured rings usually called Newton's rings change their breadth and position, when the glasses between which they appear are heated, in a manner which showed that the glasses repelled each other. M. Faye's theory of comets is based on some such repellent force. There is, however, some difficulty in presenting these phenomena to the mind in the same aspect as the molecular repulsive action of heat.

The phenomena of what is termed latent heat have been generally considered as strongly in favour of that view which regards heat either as actual matter, or, at all events, as a substantive entity, and not a motion or affection of ordinary matter.

The hypothesis of latent matter is, I venture with diffidence to think, a dangerous one—it is something like the old principle of Phlogiston, it is not tangible, visible, audible; it is, in fact, a mere subtle mental conception, and ought, I submit, only to be received on the ground of absolute necessity, the more so as these subtleties are apt to be carried on to other natural phenomena, and so they add to the hypothetical scaffolding which is seldom requisite, and should be sparingly used, even in the early stages of discovery. As an instance, I think a striking one, of the injurious effects of this, I will mention the analogous doctrine of 'invisible light;' and I do this, meaning no disrespect to its distinguished author, any more than, in discussing the doctrine of latent heat, I can be supposed, in the slightest degree, to aim at detracting from the merits of the illustrious investigators of the facts which that doctrine seeks to explain. Is not 'invisible light,' a contradiction in terms? has not light ever been regarded as that agent which affects our visual organs? Invisible light, then, is darkness, and if it exist, then is darkness light. I know it may be said, that one eye can detect light where another cannot; that a cat may see where a man cannot; that an insect may see where a cat cannot; but then it is not invisible light to those who see it: the light, or rather the object seen by the cat, may be invisible to the man, but it is visible to the cat, and, therefore, cannot abstractedly be said to be invisible. If we go further, and find an agent which affects certain substances similarly to light, but does not, as far as we are aware, affect the visual organs of any animal, then is it not an erroneous nomenclature which calls such an agent light? There are many cases in which a deviation from the once accepted meaning of words has so gradually entered into common usage as to be unavoidable, but I cannot but think that additions to such cases should as far as possible be avoided, as injurious to that precision of language which is one of the safest guards to knowledge, and from the absence of which physical science has materially suffered.

Let us now shortly examine the question of latent heat, and see whether the phenomena may not be as well, if not more satisfactorily, explained without the hypothesis of latent matter, an idea presenting many similar difficulties to that of invisible light, though more sanctioned by usage. Latent heat is supposed to be the matter of heat, associated, in a masked or dormant state, with ordinary matter, not capable of being detected by any test so long as the matter with which it is associated remains in the same physical state, but communicated to or absorbed from other bodies, when the matter with which it is associated changes its state. To take a common example: a pound or given weight of water at 172°, mixed with an equal weight of water at 32°, will acquire a mean temperature, or 102°; while water at 172°, mixed with an equal weight of ice at 32°, will be reduced to 32°. By the theory of latent heat this phenomenon is thus explained:—In the first case, that of the mixture of water with water, both the bodies being in the same physical state, no latent heat is rendered sensible, or sensible heat latent; but in the second, the ice changing its condition from the solid to the liquid state, abstracts from the liquid as much heat as it requires to maintain it in the liquid state, which it renders latent, or retains associated with itself, as long as it remains liquid, but of which heat no evidence can be afforded by any thermoscopic test.

I believe this and similar phenomena, where heat is connected with a change of state, may be explained and distinctly comprehended without recourse to the conception of latent heat, though it requires some effort of the mind to divest itself of this idea, and to view the phenomena simply in their dynamical relations. To assist us in so viewing them, let us first parallel with purely mechanical actions, certain simple effects of heat, where change of state (I mean such change as from the solid to the liquid, or liquid to the gaseous state) is not concerned. Thus, place within a receiver a bladder, and heat the air within to a higher temperature than that without it, the bladder expands; so, force the air mechanically into it by the air-pump, the bladder expands; cool the air on the outside, or remove its pressure mechanically by an exhausting pump, the bladder also expands; conversely, increase the external repellent force, either by heat or mechanical pressure, and the bladder contracts. In the mechanical effects, the force which produced the distension is derived from, and at the expense of, the mechanical power employed, as from muscular force, from gravitation, from the reacting elasticity of springs, or any similar force by which the air-pump may be worked. In the heating effects, the force is derived from the chemical action in the lamp or source of heat employed.

Let us next consider the experiment so arranged that the force, which produces expansion in the one case, produces a correlative contraction in the other: thus, if two bladders, with a connecting neck between them, be half-filled with air, as the one is made to contract by pressure the other will dilate, and vice versâ; so a bladder partly filled with cold air, and contained within another filled with hot air, expands while the space between the bladders contracts, exhibiting a mere transfer of the same amount of repulsive force, the mobility of the particles, or their mutual attraction, being the same in each body; in other words, the repulsive force acts in the direction of least resistance until equilibrium is produced; it then becomes a static or balanced, instead of a dynamic or motive force.

Let us now consider the case where a solid is to be changed to a liquid, or a liquid to a gas; here a much greater amount of heat or repulsive force is required, on account of the cohesion of the particles to be separated. In order to separate the particles of the solid, precisely as much force must be parted with by the warmer liquid body as keeps an equal quantity of it in its liquid state; it is, indeed, only with a more striking line of demarcation, the case of the hot and cold bladder--a part of the repellent power of the hot particles is transferred to the cold particles, and separates them in their turn, but the antagonist force of cohesion or aggregation necessary to be overcome, being in this case much stronger, requires and exhausts an exactly proportionate amount of repellent force mechanically to overcome it; hence the different effect on a body such as the common thermometer, the expanding liquid of which does not undergo a similar change of state. Thus, in the example above given, of the mixture of cold with hot water, the hot and cold water and the mercury of the thermometer being all in a liquid state before, and remaining so after contact, the resulting temperature is an exact mean; the hot water contracts to a certain extent, the cold water expands to the same extent, and the thermometer either sinks or rises the same number of degrees, accordingly as it had been previously immersed in the cold or in the hot solution, its mercury gaining or losing an equivalent of repellent force. In the second instance, viz. the mixture of ice with hot water, the substance we use as an indicator, i.e. mercury, does not undergo the same physical change as do those whose relations of volume we are examining. The force—viewing heat simply as mechanical force—which is employed in loosening or tearing asunder the particles of the solid ice, is abstracted from the liquid water, and from the liquid mercury of the thermometer, and in proportion as this force meets with a greater resistance in separating the particles of a solid than of a liquid, so the bodies which yield the force suffer proportionately a greater contraction.

If we compare the action of heat on the two substances, water and mercury alone, and throw out of our consideration the ice, we shall be able to apply the same view: thus, if a given source of heat be applied to water containing a mercurial thermometer, both the water and mercury gradually expand, but in different degrees; at a certain point the attractive force of the molecules of the water is so far overcome that the water becomes vapour. At this point, the heat or force, meeting with much less resistance from the attraction of the particles of steam than from those of the mercury, expends itself upon the former; the mercury does not further expand, or expands in an infinitesimally small degree, and the steam expands greatly. As soon as this arrives at a point where circumambient pressure causes its resistance to further expansion to be equal to the resistance to expansion in the mercury of the thermometer, the latter again rises, and so both go on expanding in an inverse ratio to their molecular attractive force. If the circumambient pressure be increased, as by confining the water at the commencement of the experiment within a less expansible body than itself, such as a metallic chamber, then the mercury of the thermometer continues to rise; and if the experiment were continued, the water being confined and not the mercury, until we have arrived at a degree of repulsive force which is able to overcome the cohesive power of the mercury, so that this expands into vapour, then we get the converse effect; the force expends itself upon the mercury, which expands indefinitely, as the water did in the first case, and the water does not expand at all.

Another very usual mode of regarding the subject may embarrass at first sight, but a little consideration will show that it is explicable by the same doctrine. Water which has ice floating in it will give, when measured by the thermometer, the same temperature as the ice; i.e. both the water and ice contract the mercury of the thermometer to the point conventionally marked as 32°. It may be said, how is this reconcileable with the dynamical doctrine, for according to that the solid should take from the mercury of the thermometer more repulsive power than the liquid; consequently, the ice should contract the mercury more than the water?

My answer is, that in the proposition as thus stated, the quantities of the water, ice, and mercury are not taken into consideration, and hence a necessary dynamical element is neglected: if the element of quantity be included, this objection will not apply. Let the thermometer, for instance, contain $13\frac{1}{2}$ oz. of mercury, and stand at 100°; if placed in contact with an unlimited quantity of ice at 32°, the mercury will sink to 32°. If the same thermometer be immersed in an unlimited quantity of water at 32°, the mercury sinks also to 32°; not absolutely perhaps, because, however great the quantity of water or ice, it will be somewhat raised in temperature by the warmer mercury. This elevation of temperature above 32°, will be smaller in proportion as the quantity of water or ice is larger than the quantity of mercury; and, as we know of no intermediate state between ice and water, the contact of a thermometer at a temperature above the freezing point with any quantity of ice exactly at the freezing point would, theoretically speaking, liquefy the whole, provided it had sufficient time; for as every portion of that ice would in time have its temperature raised by the contact of the warmer body, and as any elevation of temperature above the freezing point liquefies ice, every portion should be liquefied. Practically speaking, however, in both cases, that of the water and of the ice, when the quantity is indefinitely great the thermometer falls to 32°.

Now place the same thermometer at 100°, successively in one oz. of water at 32°, and in one of ice at 32°; we shall find in the former case it will be lowered only to 54°, and in the latter to 32°; apply to this the doctrine of repulsive force, and we get a satisfactory explanation.

In the first case, the quantity both of ice and water being indefinitely great in respect to the mercury, each reduces it to its own temperature, viz. 32°, and the ice cannot reduce the mercury below 32°, because the latter would receive back repulsive power from the newly formed water, and this would become ice; in the second case, where the quantities are limited, the mercury does lose more repulsive power by the ice than by the water, and the observations made in reference to the first illustration apply.

The above doctrine is beautifully instanced in the experiment of Thilorier, by which carbonic acid is solidified. Carbonic acid gas, retained in a strong vessel under great pressure, is allowed to escape from a small orifice; the sudden expansion requires so great a supply of force, that in furnishing the demands of the expanding gas certain other portions of the gas contract to such an extent as to solidify: thus, we have reciprocal expansion and contraction going on in one and the same substance, the time being too limited for the whole to assume a uniform temperature, or in other words, a uniform extent of expansion.

It has been observed with reference to heat thus viewed, that it would be as correct to say, that heat is absorbed, or cold produced by motion, as that heat is produced by it. This difficulty ceases when the mind has been accustomed to regard heat and cold as themselves, motion; i.e. as correlative expansions and contractions, each being evidenced by relation, and being inconceivable as an abstraction.

For instance, if the piston of an air-pump be drawn down by a weight, cold is produced in the receiver. It may be here said that a mechanical force, and the motion consequent upon it, produces cold; but heat is produced on the opposite side of the piston, if a receiver be adapted so as to retain the compressed air. Assuming them to be equivalent to each other, the force of the falling weight would be expressed by the heat of friction of the piston against its tube, and by the tension or power of reaction of the compressed against the dilated air. If the heat due to compression be made to perform mechanical work, it would *pro tanto* be consumed, and could not restore the temperature to the dilated air; but if it perform no work, no heat is lost. Mr. Joule has experimentally proved this proposition.

In commencing the subject of heat, I asked my reader to put out of consideration the sensations which heat produces in our own bodies: I did this because these sensations are likely to deceive, and have deceived many as to the nature of heat. These sensations are themselves occasioned by similar expansions to those which we have been considering; the liquids of the body are expanded, i. e. rendered less viscid by heat, and from their more ready flow, we obtain the sensation of agreeable warmth. By a greater degree of heat, their expansion becomes too great, giving rise to a sense of pain, and if pushed to extremity, as with the heat which produces a burn, the liquids of the body are dissipated in vapour, and an injury or destruction of the organic structure takes place. A similar though converse effect may be produced by intense cold; the application of frozen mercury to the animal body produces a burn similar to that produced by great heat, and accompanied with a similar sensation.

Doubtless other actions than those above mentioned interfere in producing the sensations of heat and cold; but I think it will be seen that these will not affect the arguments as to the nature of heat. The phenomenal effects will be found unaltered: heat will still be found to be expansion, cold to be contraction; and the expansion and contraction are, as with the two bladders of air, correlative—i.e. we cannot expand one body, A, without contracting some other body, B; we cannot contract A without expanding B, assuming that we view the bodies with relation to heat alone, and suppose no other force to be manifested.

I have said that there are a few exceptions as to heat being always manifested by an expansion of matter. One class of these exceptions is only apparent: moist clay, animal or vegetable fibre, and other substances of a mixed nature, which contain matter of diverse character, some of which is more and some less volatile, i.e. expansible, are contracted on the application of heat: this arises from the more volatile matter being dissipated in the form of vapour or gas; and the interstices of the less volatile being thus emptied, the latter contracts by its own cohesive attraction, giving thus a primâ facie appearance of contraction by heat. The pyrometer of Wedgwood is explicable on this principle.

The second class of exceptions, though much more limited in extent, is less easily explained. Water, fused bismuth, and probably some other substances (though the fact as to them is not clearly established), expand as they approach very near to the freezing or solidifying point. The most probable explanation of these exceptions is, that at the point of maximum density the molecules of these bodies assume a polar or crystalline condition; that by the particles being thus arranged in linear directions like chevaux de frise, interstitial spaces are left, containing matter of less density, so that the specific density of the whole mass is diminished.

Some recent experiments of Dr. Tyndall on the physical properties of ice seem to favour this view. When a sunbeam, concentrated by a lens, is allowed to fall on a piece of apparently homogeneous ice the path of the rays is instantly studded with numerous luminous spots like minute air bubbles, and the planes of freezing are made manifest by these and by small fissures. Stars or flower-like figures of six petals appear parallel to the planes of freezing, and seemingly spreading out from a central bubble. These flowers are formed of water. When the ice is melted in warm water no air is given off from the bubbles, so they seem to be vacuous; it is, however, possible that extremely minute particles of air sufficient to form foci for the melting points of ice might be dissolved by the water as soon as they came in contact with it. Be this as it may, the existence of these points throughout the ice, where it gives way to the heat of the solar beam, if it does not prove actual vacuous or aeriform spaces to exist in ice, proves that it is not homogeneous, that its structure is probably definitely crystalline, and that the matter composing it is in different degrees of aggregation, so that its mean specific gravity might well be less than that of water.

We cannot examine piecemeal the ultimate structure of matter, but in addition to the fact that the bodies which evince this peculiarity are bodies which, when solidified, exhibit a very marked crystalline character, there are experiments which show that water between the point of maximum density and its point of solidification polarises light circularly; showing, if these experiments be correct, a structural alteration in water, and one analogous to that possessed by certain crystalline solids, and to that possessed by water itself, where it is forcibly made to assume a polarised condition by the influence of magnetism.

The accuracy of these results has, however, been doubted, and the experiments have not succeeded when repeated by very experienced hands. Whether this be so or not, and whether the above explanation of the exception to the otherwise invariable effect of expansion by heat be or be not regarded as admissible, must be left to the judgement and experience of each individual who thinks upon the subject; at all events, no theory of heat yet proposed removes the difficulty, and therefore it equally opposes every other view of the phenomena of heat, as it does that which I have here considered, and which regards heat as communicable expansive force.

As certain bodies expand in freezing, and indeed, under some circumstances, before they arrive at the temperature at which they solidify, we get the apparent anomaly that the motion or mechanical force generated by heat or change of temperature is reversed in direction when we arrive at the point of change from the solid to the liquid state. Thus a piece of ice at the temperature of Zero, Fahrenheit, would expand by heat, and produce a mechanical force by such expansion until it arrives at 32°; but then by an increment of heat it contracts, and if the first expansion had moved a piston upwards, the subsequent contraction would bring it back to a certain extent, or move it downwards, an apparent negation of the force of heat.

Again with water above 40°, i.e. above its point of maximum density, a progressive increment of cold or decrement of heat would produce contraction to a certain point, and then expansion or a mechanical force in an opposite direction.

Thus not only heat or the expansive force given to other bodies by a body cooling would be given out by water freezing, but also the force due to the expansion in the body itself, and force would thus seem to be got out of nothing: but if water in a confined space be gradually cooled, the expansion attendant on its cooling as it approaches the freezing point would occasion pressure amongst its particles, and thence tend to antagonise the force of dilatation produced in them by cooling, or to resist their tendency to freeze; or in other words the pressure would tend to liquefaction, and conversely to the usual effect of pressure, produce cold instead of heat, and thus neutralise some of the heat yielded by the cooling body. Hence we find that it requires a lower temperature to freeze water under pressure than when exempt from it, or that the freezing point is lowered as the pressure increases for bodies which expand in freezing—an effect first predicted by Mr. J. Thompson, and experimentally verified by Mr. W. Thompson; while as shown by M. Bunsen, the converse effect takes place with bodies which contract in freezing. Here the pressure cooperates with the effects of cold, both tending to approximate the particles, and such substances solidify at a higher temperature in proportion as the pressure is greater; so that we might expect a body of this class, which under the ordinary pressure of the air is at a temperature just above its freezing point, to solidify by being submitted to pressure alone, the temperature being kept constant.

A similar class of exception to the general effect of heat in expanding bodies is presented by vulcanised caoutchouc. This has been observed by Mr. Gough, and, indeed, was pointed out to me many years ago by Mr. Brockedon to be heated when stretched, and cooled when unstretched.

Mr. Joule finds that its specific gravity is lower when stretched than when unstretched, and that when heated in its stretched state it shortens, presenting in this particular condition a similar series of relations to those which are presented by water near or at its freezing point.

With the exception of this class of phenomena, which offer difficulties to any theory which has been proposed, the general phenomena of heat may, I believe, be explained upon a purely dynamical view, and more satisfactorily than by having recourse to the hypothesis of latent matter. Many, however, of the phenomena of heat are involved in much mystery, particularly those connected with specific heat or that relative proportion of heat which equal weights of different bodies require to raise them from a given temperature to another given temperature, which appear to depend in some way hitherto inexplicable upon the molecular constitution of different bodies.

The view of heat which I have taken, viz. to regard it simply as a communicable molecular repulsive force, is supported by many of the phenomena to which the term specific or relative heat is applied; for example, bodies as they increase in temperature increase in specific heat. The ratio of this increase in specific heat is greater with solids than with liquids, although the latter are more dilatable; an effect probably depending upon the commencement of fusion. Again, those metals whose rate of expansion increases most rapidly when they are heated, increase most in specific heat; and their specific heat is reduced by percussion, which, by approximating their particles, makes them specifically more dense. When, however, we examine substances of very different physical characters, we find that their specific heats have no relation to their density or rate of expansion by heat; their differences of specific heat must depend upon their intimate molecular constitution in a manner accounted for (as far as I am aware) by no theory of heat hitherto proposed.

In the greater number, probably in all solids and liquids, the expansion by heat is relatively greater as the temperature is higher; or, preserving the view of expansion and contraction, if two equal portions of the same substance be juxtaposed at different temperatures, the hotter portion will contract a little more than the colder will expand; from this fact, viz. that the coefficient of expansion increases in a given body with the temperature, and from other considerations, Dr. Wood has argued, with much apparent reason, that the nearer the particles of bodies are to each other, the less they require to move to produce a given expansion or contraction in those of another body. His mode of reasoning, if I rightly conceive it, may be concisely put as follows:—

As bodies contract by cold, it is clear that, in a given body, the lower the temperature the nearer are the particles; and, as the coefficient of expansion increases with the temperature, the lower the temperature of the substance be, the less the particles require to move, or approach to or recede from each other, so as to compensate the correlative recession or approach of the particles in a hotter portion of the same substance, that is in another portion of the same substance in which the particles are more distant from each other. The amount of approximation or recession of the particles of a body, in other words, its change of bulk by a given change of temperature, being thus in a given substance an index of the relative proximity of its particles, may it not be so of all bodies? The proposition is very ingeniously argued by Dr. Wood, but the argument is based upon certain hypotheses as to the sizes and distances of atoms, which must be admitted as postulates by those who adopt his conclusions. Dr.

Wood seeks by means of this theory to explain the heat produced by chemical combination, and I shall endeavour to give a sketch of his mode of reasoning when I arrive at that part of my subject.

Although the comparative effects of specific heat may not be satisfactorily explicable by any known theory, the absolute effect of heat upon each separate substance is simply expansion, but when bodies differing in their physical characters are used, the rate of expansion varies, if measured by the correlative contractions exhibited by the substances producing it. Though I am obliged, in order to be intelligible, to talk of heat as an entity, and of its conduction, radiation, &c., yet these expressions are, in fact, inconsistent with the dynamic theory which regards heat as motion and nothing else; thus conduction would be simply a progressive dilatation or motion of the particles of the conducting substance, radiation an undulation or motion of the particles of the medium through which the heat is said to be transmitted, &c.; and it is a strong argument in favour of this theory, that for every diversity in the physical character of bodies, and for every change in the structure and arrangement of particles of the same body, a change is apparent in the thermal effects. Thus gold conducts heat, or transmits the motion called heat, more readily than copper, copper than iron, iron than lead, and lead than porcelain, &c.

So when the structure of a substance is not homogeneous, we have a change in the conduction of different parts dependent upon the structure. This is beautifully shown with bodies whose structure is symmetrically arranged, as in crystals. Senarmont has shown that crystals conduct heat differently in different directions with reference to the axis of symmetry, but definitely in definite directions. His mode of experimenting is as follows:—A plate of the crystal is cut in a direction, for one set of experiments parallel, and for another at right angles to the axis; a tube of platinum is inserted through the centre of the plate, and bent at one extremity, so as to be capable of being heated by a lamp without the heat which radiates from the lamp affecting the crystal; the surfaces or bases of the plate of crystal are covered with wax. When the platinum is heated, the direction of the heat conducted by the crystal is made known by the melting of the wax, and a curved line is visible at the juncture of the solid and liquid wax. This curve, with homogeneous substances, as glass or zinc, is a circle; it is also a circle on plates of calc spar cut perpendicular to the axis of symmetry; but on plates cut parallel to the axis of symmetry, and having their plane perpendicular to one of the faces of the primitive rhombohedron, the curves are well-defined ellipses, having their longer axes in the direction of the axis of symmetry, showing that this axis is a direction of greater conductibility. From experiments of this character the inference is drawn, that 'in media constituted like crystals of the rhombohedral system, the conducting power varies in such a manner, that, supposing a centre of heat to exist within them, and the medium to be indefinitely extended in all directions, the isothermal surfaces are concentric ellipsoids of revolution round the axis of symmetry, or at least surfaces differing but little therefrom.'

Knoblauch has further shown, that radiant heat is absorbed in different degrees, according as its direction is parallel or perpendicular to the axis of a crystal.

If we select a substance of a different but also of a definite structure, such as wood, we find that heat progresses through it with more or less rapidity, according to its direction with reference to the fibre of the wood: thus Decandolle and De la Rive found that the conduction was better in a direction parallel to the fibre than in one transverse to it; and Dr. Tyndall has added the fact, that the conduction is better in a direction transverse to the fibre and layers of the wood than when transverse to the fibre but parallel to the layers, though in both these directions the conduction is inferior to that following the direction of the fibre. Thus, in the three possible directions in which the structure of wood may be contemplated, we have three different degrees of progression for heat.

In the above examples we see, as we shall see farther on with reference to all the so-called imponderables, that the phenomena depend upon the molecular structure of the matter affected; and although these facts are not absolutely inconsistent with the theory which supposes them to be fluids or entities, it will, I think, be found to be far more consistent with that which views them as motion. Heat, which we are at present considering, cannot be insulated: we cannot remove the heat from a substance and retain it as heat; we can only transmit it to another substance, either as heat or as some other mode of force. We only know certain changes of matter, for which changes heat is a generic name; the *thing* heat is unknown.

Heat having been shown to be a force capable of producing *motion*, and motion to be capable of producing the other modes of force, it necessarily follows that heat is capable, mediately, of producing them: I will, therefore, content myself with enquiring how far heat is capable of immediately producing the other modes of force. It will immediately produce *electricity*, as shown in the beautiful experiments of Seebeck, one of which I have already cited, which experiments proved, that when dissimilar metals are made to touch, or are soldered together and heated at the point of contact, a current of electricity flows through the metals having a definite direction according to the metals employed, which current continues as long as an increasing temperature is gradually pervading the metals, ceases when the temperature is stationary, and flows in the contrary direction with the decrement of temperature.

Another class of phenomena which have been generally attributed to the effects of radiant heat, and to which, from this belief, the term thermography has been applied, may also, in their turn, be made to exhibit electrical effects—effects here of Franklinic or static electricity, as Seebeck's experiments showed effects of voltaic or dynamic electricity.

If polished discs of dissimilar metals—say, zinc and copper—be brought into close proximity, and kept there for some time, and either of them has irregularities upon its surface, a superficial outline of these irregularities is traceable upon the other disc, and vice versâ. Many theories have been framed to account for this phenomenon, but whether it be due or not to thermic radiations, the relative temperature of the discs, their relative capacities and conducting and radiating powers for heat, undoubtedly influence the phenomena.

Now, if two such discs in close proximity be connected with a delicate electroscope, and then suddenly separated, the electroscope is affected, showing that the reciprocal radiation from surface to surface has produced electrical force. I cite this experiment in treating of heat as an initial force, because at present the probabilities are in favour of thermic radiation producing the phenomenon. The origin of these so-called thermographic effects is, however, a question open to much doubt, and needs much further experiment. When I first published the experiment which showed that the mere approximation of discs of dissimilar metals would give rise to electrical effects, I mentioned that I considered the fact of the superficial change upon the surface of metals in proximity, and, *à fortiori*, in contact, would explain the developement of electricity in Volta's original contact experiment, without having recourse to the contact theory, i. e. a theory which supposes a force to be produced by mere contact of dissimilar metals without any molecular or chemical change. I have seen nothing to alter this view. Mr. Gassiot has repeated and verified my experiment with more delicate apparatus and under more unexceptionable circumstances; and without saying that radiant heat is the initial force in this case, we have evidence, by the superficial change which takes place in bodies closely approximated, that some molecular change is taking place, some force is called into action by their proximity, which produces changes in matter as it expends, or rather transmits itself; and, therefore, is not a force without molecular change, as the supposed contact force would be. The force in this, as in all other cases, is not created, but developed by the action of matter on matter, and not annihilated, as it is shown by this experiment to be convertible into another mode of force.

To say that heat will produce *light*, is to assert a fact apparently familiar to everyone, but there may be some reason to doubt whether the expression to produce light is correct in this particular application; the relation between heat and light is not analogous to the correlation between these and the other four affections of matter. Heat and light appear to be rather modifications of the same force than distinct forces mutually dependent. The modes of action of radiant heat and of light are so similar, both being subject to the same laws of reflection, refraction, double refraction, and polarisation, that their difference appears to exist more in the manner in which they affect our senses than in our mental conception of them.

The experiments of Melloni, which have mainly contributed to demonstrate the analogies of heat and light, afford a beautiful instance of the assistance which the progress of one branch of physical science renders to that of another. The discoveries of Œrsted and Seebeck led to the construction of an instrument for measuring temperature, incomparably more delicate than any previously known. To distinguish it from the ordinary thermometer, this instrument is called the *thermomultiplier*. It consists of a series of small bars of bismuth and antimony, forming one zigzag chain of alternations arranged parallel to each other, in the shape of a cylinder or prism; so that the points of junction, which are soldered, shall be all exposed at the bases of the cylinder: the two extremities of this series are united to a galvanometer—that is, a flat coil of wire surrounding a freely-suspended magnetic needle, the direction of which is parallel to the convolutions of the wire. When radiant heat impinges upon the soldered ends of the multiplier, a thermo-electric current is induced in each pair; and, as all these currents tend to circulate in the same direction, the energy of the whole is increased by the cooperating forces: this current, traversing the helix of the galvanometer, deflects the needle from parallelism by virtue of the electro-magnetic tangential force, and the degree of this deflection serves as the index of the temperature.

Bodies examined by these means show a remarkable difference between their transcalescence, or power of transmitting heat, and their transparency: thus, perfectly transparent alum arrests more heat than quartz so dark coloured as to be opaque; and alum coupled with green glass Melloni found was capable of transmitting a beam of brilliant light, while, with the most delicate thermoscope, he could detect no indications of transmitted heat: on the other hand, rock-salt, the most transcalescent body known, may be covered with soot until perfectly opaque, and yet be found capable of transmitting a considerable quantity of heat. Radiant heat, when transmitted through a prism of rock-salt, is found to be unequally refracted, as is the case with light; and the rays of heat thus elongated into what is, for the sake of analogy, called a spectrum, are found to possess similar properties to the primary or coloured rays of light. Thus rock-salt is to heat what colourless glass is to light; it transmits heat of all degrees of refrangibility: alum is to heat as red glass to light; it transmits the least, and stops the most refrangible rays; and rock-salt covered with soot represents blue glass, transmitting the most, and stopping the least refrangible rays.

Certain bodies, again, reflect heat of different refrangibility: thus, paper, snow, and lime, although perfectly white—that is, reflecting light of all degrees of refrangibility, reflect heat only of certain degrees; while metals, which are coloured bodies—that is, bodies which reflect light only of certain degrees of refrangibility—reflect heat of all degrees. Radiant heat incident upon substances which doubly refract light is doubly refracted; and the emergent rays are polarised in planes at right angles to each other, as is the case with light.

The relation of radiation to absorption also holds good with light as with heat: with the latter it has been long known that the radiating power of different substances is directly proportional to their absorptive and inverse to their reflective power; or rather, that the sum of the heat radiated and reflected is a constant quantity. Further, it has been shown by Mr. Balfour Stewart, that the absorption of heat is proportional to its radiation, both as to quality and quantity.

Light presents us with similar relations. Coloured glass, when heated so as to be luminous, emits the same light which at ordinary temperatures it absorbs: thus red glass gives out or radiates a greenish light, and green glass a red tint.

The flame of substances containing sodium yields a yellow light of such purity that other colours exposed to it appear black—a phenomenon shown by the familiar experiment of exposing a picture of bright colours, other than yellow, to the flame of spirits of wine with which common salt is mixed: the picture loses its colours, and appears to be black and white. When the prismatic spectrum of such a flame is examined, it is found to exhibit two bright yellow lines at a certain fixed position. If a source of light be employed which gives no lines in its spectrum, and the light of this incandescent substance be made to pass through the sodium flame, two dark lines will appear in the spectrum precisely coincident in position with the yellow lines which were given by the sodium flame itself. The same relation of absorption to radiation is therefore shown here: the substance absorbs that light which it yields when it is itself the source of light. The same is true of other substances, the spectra of which exhibit respectively lines of peculiar colour and position. Now, the solar prismatic spectrum is traversed by a great number of dark lines; and Kirchhoff has deduced from considerations such as those which I have shortly stated, that these dark lines in the solar spectrum are due to metals existing in an atmosphere around the sun, which absorb the light from a central incandescent nucleus, each metal absorbing that light which would appear as a bright line or lines in the spectrum produced by its own light.

By comparing the positions of the bright lines in the spectra of various metals with those of the dark lines in the solar spectrum, several of them are found to be in identically the same place: hence it is inferred, and the inference seems reasonable, that the metals which show luminous lines in their spectra, identical in position with dark lines in the solar spectrum, exist in the sun, and are diffused in a gaseous state in its atmosphere. It does not seem to me necessary to this conclusion to assume that the sun is a solid mass of incandescent matter: it may well be that what we term the photosphere or luminous envelope of the sun has surrounding it a more diffuse atmosphere containing vaporised metals, and that the mass of the sun itself may be in a different state, and not necessarily at an incandescent temperature; indeed, the protuberances and red light seen at the period of total eclipses afford some evidence of an atmosphere exterior to the photosphere. It would, however, be out of place here to speculate on these subjects: the point which concerns us is the analogy of heat and light, which these discoveries illustrate. Kirchhoff has carried the analogy farther by showing that a plate of tourmaline absorbs the polarised ray which when heated it radiates. Thus, the phenomena of light are imitated closely by those of radiant heat; and the same theory which is considered most plausibly to account for the phenomena of the one, will necessarily be applied to the other agent, and in each case molecular change is accompanied by a change in the phenomenal effects.

In certain cases heat appears to become partially converted into light, by changing the matter affected by heat: thus gas may be heated to a very high point without producing light, or producing it to a very slight degree; but the introduction of solid matter—for instance, the metal platinum into the highly-heated gas—instantly exhibits light. Whether the heat is converted into light, or whether it is concentrated and increased in intensity by the solid matter so as to become visible, may be open to some doubt : the fact of solid matter, when ignited by the oxyhydrogen jet, decomposing water, as will be presently explained, would seem to indicate that the heat was rendered more intense by condensation in the solid matter, as water is in this case decomposed by a heated body, which body has itself been heated by the combining elements of water. The apparent effect, however, of the introduction of solid incombustible matter into heated gas, is a conversion of heat into light.

Dr. Tyndall, by passing light from the voltaic arc through solutions of iodine, separates the invisible rays of heat from the luminous rays, and then reproduces light by receiving the former on platinum foil.

If we concentrate into a focus by a large lens a dim light, we increase the intensity of the light. Now if a heated body be taken, which, to the unassisted eye, has just ceased to be visible, it seems probable that by collecting and condensing by a lens the different rays which have so ceased to be visible, light would reappear at the focus. The experiment is, for reasons obvious to those acquainted with optics. a difficult one, and, to be conclusive, should be made on a large scale, and with a very perfect lens of large diameter and short focus. I have obtained an approximation to the result in the following manner:—In a dark room a platinum wire is brought just to the point of visible ignition by a constant voltaic battery; it is then viewed, at a short distance, through an opera-glass of large aperture applied to one eye, the other being kept open. The wire will be distinctly visible to that eye which regards it through the opera-glass, and at the same time totally invisible to the other and naked eye. It may be said with some justice that such experiments prove little more than the fact already known, viz. that by increasing the intensity of heat, light is produced; they however exhibit this effect in a more striking form, as bearing on the relations of heat and light.

With regard to *chemical affinity* and *magnetism*, perhaps the only method by which in strictness the force of heat may be said to produce them is through the medium of electricity, the thermo-electrical current, produced, as before described, by heating dissimilar metals, being capable of deflecting the magnet, of magnetising iron, and exhibiting the other magnetic effects, and also of forming and decomposing chemical compounds, and this in proportion to the progression of heat: this has not, indeed, as yet been proved to bear a measurable quantitative relation to the other forces thus produced by it, because so little of the heat is utilised or converted into electricity, much being dissipated, without change, in the form of heat.

Heat, however, directly affects and modifies both

the magnet and chemical compounds; the union of certain chemical substances is induced by heat, as, for instance, the formation of water by the union of oxygen and hydrogen gases: in other cases this union is facilitated by heat, and in many instances, as in ammonia and its salts, it is weakened or antagonised. In many of these cases, however, the force of heat seems more a determining than a producing influence; yet to be this, it must have an immediate relation with the force whose reaction it determines: thus, although gunpowder, touched with an ignited wire, subsequently carries on its own combustion or chemical combination, independently of the original source of heat, yet the chemical affinities of the first portion touched must be exalted by, and at the cost of, the heat of the wire; for to disturb even an unstable equilibrium requires a force in direct relation with those which maintain equilibrium.

Since the first edition of this essay was published, I have communicated to the Royal Society some experiments by which an important exception to the general effect of heat on chemical affinity is removed, and the results of which induce a hope that a generalised relation will ultimately be established between heat, chemical affinity, and physical attraction. I find that if a substance capable of supporting an intense heat, and incapable of being acted upon by water or either of its elements—such, for instance, as platinum, or iridium — be raised to a high point of ignition and then immersed in water, bubbles of permanent gas ascend from it, which on examination are found to consist of mixed oxygen and hydrogen in the proportions in which they form water. The temperature at which this is effected is, according to Dr. Robinson, who has since written a valuable paper on the subject, $= 2386°$. Now, when mixed oxygen and hydrogen are exposed to a temperature of about 800°, they combine and form water; heat therefore appears to act differently upon these elements according to its intensity, in one case producing composition, in the other decomposition. No satisfactory means of reconciling this apparent anomaly have been pointed out: the best approximation to a hypothesis which I can frame is by assuming that the constituent molecules of water are, below a certain temperature, in a state of stable equilibrium; that the molecules of mixed or oxyhydrogen gas are, above a certain temperature, also in a state of stable equilibrium, but of an opposite character; while below this latter temperature the molecules of mixed gas are in a state of unstable equilibrium, somewhat similar to that of the fulminates or similar bodies, in which a slight derangement subverts the nicely-balanced forces.

   If, for instance, we suppose four molecules,

A, B, C, D, to be in a balanced state of equilibrium between attracting and repelling forces, the application of a repulsive force between B and C, though it may still farther separate B and C, will approximate B to A and C to D, and may bring them respectively within the range of attractive force; or, supposing the repulsive force to be in the centre of an indefinite sphere of particles, all these, excepting those immediately acted on by the force, will be approximated, and having from attraction assumed a state of stable equilibrium, they will retain this, because the repulsive force divided by the mass is not capable of overcoming it. But if the repulsive force be increased in quantity and of sufficient intensity, then the attractive force of all the molecules may be overcome, and decomposition ensue. Thus, water or steam below a certain temperature, and mixed gas above a certain temperature, may be supposed to be in the state of stable equilibrium, whilst below this limiting temperature, the equilibrium of oxyhydrogen gas is unstable.

This, it must be confessed, is but a crude mode of explaining the phenomena, and requires the assumption, that the particles of a gas exercise an attraction for each other as do the particles of a solid, though different in degree, perhaps in kind. Whether this be so or not, there can be no doubt that both gases and solids expand or contract according to the inverse contraction or expansion of other neighbouring bodies, and so far resemble each other in their relations to heat and cold. The extent to which such expansion or contraction can be carried, seems to be limited only by the correlative state of other bodies; these, again, by others, and so on, as far as we may judge, throughout the universe.

Adopting the explanation above given of the decomposition of water by heat, heat would have the same relation to chemical affinity as it has to physical attraction; its immediate tendency is antagonistic to both, and it is only by a secondary action that chemical affinity is apparently promoted by heat. This hypothesis would account for heat promoting changes of the equilibrium of chemical affinity among mixed compound substances, by decomposing certain compounds and separating elementary constituents whose affinity is greater, when they are brought within the sphere of attraction for the substance with which they are mixed, than for those with which they were originally chemically united: thus an intense heat being applied to a mixture of chlorine and the vapour of water, occasions the production of muriatic acid, liberating oxygen.

Carrying out this view, it would appear that a sufficient intensity of heat might yield indefinite powers of decomposition; and there seems some probability of bodies now supposed to be elementary, being decomposed or resolved into further elements by the application of heat of sufficient intensity; or, reasoning conversely, it may fairly be anticipated that bodies, which will not enter into combination at a certain temperature, will enter into combination if their temperature be lowered, and that thus new compounds may be formed by a proper disposition of their constituents when exposed to an extremely low temperature, and the more so if compression be also employed.

In considering the effect of heat as a mechanical force, it would be expected, *à priori*, and independently of any theory of heat which may be adopted, that a given amount of heat acting on a given material must produce a given amount of motive power; and the next question which occurs to the mind is, whether the same amount of heat would produce the same amount of mechanical power, whatever be the material acted on or affected by the heat. I will endeavour to reason this out on the view of heat which I have advocated. Heat has been considered in this essay as itself motion or mechanical power, and quantity of heat as measured by motion. Thus, if by a given contraction of a body (say mercury) air within a cylinder having a moveable piston be expanded, the piston moves, and in this case the expansion or motion of the material (say iron) of the cylinder itself and of the air surrounding it is commonly neglected. As the air dilates it becomes colder; in other words, by undergoing expansion itself, it loses its power of making neighbouring bodies expand; but if the piston be forcibly kept down, the expansive power due to the mercury continues to communicate itself to the iron and to the surrounding air, which become hotter than they would if the piston had given way.

Now, in the above case, if the air be confined and its volume unchanged, will the expansion of the iron, assuming that it can be utilised, produce an exactly equivalent mechanical effect to that which the expansion of the air would produce if the heat be entirely confined to it?

Assuming that (with the exception of bodies which expand in freezing, where, through a limited range of temperature, the converse effects obtain) whenever a body is compressed it is heated, i.e. it expands neighbouring substances; whenever it is dilated or increased in volume it is cooled, i.e. it contracts neighbouring substances—the conclusion appears to me inevitable that the mechanical power produced by heat will be definite, or the same for a given amount and intensity of heat, whatever be the substance acted on.

Thus, let A be a definite source of heat, say a pound of mercury at the temperature of 400°; let B be another equal and similar source of heat:

suppose A be employed to raise a piston by the dilatation of air, and B to raise another piston by the dilatation of the vapour of water. Imagine the pistons attached to a beam, so that they oppose each other's action, and thus represent a sort of calorific balance. If A being applied to air could conquer B, which is applied to water, it would depress or throw back the piston of the latter, and, by compressing the vapour, occasion an increase of temperature; this, in its turn, would raise the temperature of the source of heat, so that we should have the anomaly that a pound of mercury at 400° could heat another pound of mercury at 400° to 401°, or to some point higher than its original temperature, and this without any adventitious aid: it will be obvious that this is impossible, at least contradictory to the whole range of our experience.

The above experiment is ideal, and stated for the object of giving a more precise form to the reasoning; to bring the idea more prominently into relief, all statements as to quantities, specific heats, &c., so as to yield comparative results for given materials, are omitted. The argument may be thus stated in another form, viz. that by no mechanical appliance or difference of material acted on can a given source of heat be made to produce more heat than it originally possessed; and that, if all be converted into mechanical power, an excess cannot be supposed, for that could be converted into a surplus of heat, and be a creation of force; and a deficit cannot be supposed, for that would be annihilation of force. I cannot, however, see how the theoretical conception could be verified by experiment; the enormous weights and the complex mechanical contrivances requisite to give the measure of power yielded by matter in its less dilatable forms, would be far beyond our present experimental resources. It would also be difficult to prevent the interference of molecular attractions, inertia, &c., the overcoming of which expends a part of the mechanical power generated, but which could hardly be made to appear in the result. We could not, for instance, practically realise the above conception by the construction of a machine which should act by the expansion and contraction of a bar of iron, and produce a power equal to that of a steam engine, supplied with an equal quantity of heat.

Carnot, who wrote in 1824 an essay on the motive power of heat, regarded the mechanical power produced by heat as resulting from a transfer of heat from one point to another, without any ultimate loss of heat. Thus, in the action of an ordinary steam engine, the heat from the furnace having expanded the water of the boiler and raised the piston, a mechanical motion is produced; but this cannot be continued without the removal of the heat, or the contraction of the expanded water. This is done by the condenser, and the piston descends. But then we have apparently transferred the heat from the furnace to the condenser, and in the transfer effected mechanical motion.

Should the mechanical motion produced by heat be considered as the effect of a simple transference of heat from one point to another, or as the result of a conversion of heat into the mechanical force of which this motion is the result? This question leads to the following: does the heat which generates the mechanical power return to the thermal machine as heat, or is it conveyed away by the work performed?

If a definite quantity of air be heated it is expanded, and by its expansion it cools or loses some of its power of communicating heat to neighbouring bodies. That which we should have called heat if the expansion of the air had been prevented, we call mechanical effect, or may view as converted into mechanical effect ceasing to be heat; but, throwing out of the question nervous sensation, this expansion or mechanical effect is all the evidence we have of heat, for if the air is allowed to expand freely, this expansion becomes the index of the heat; if the air be confined, the expansion of the matter of the vessel confining it, or of the mercury of a thermometer in contact with it, &c., are the indices of the heat.

If, again, the air which has been expanded be, by mechanical pressure or by other means, restored to its original bulk, it is capable of heating or expanding other substances to a degree to which it would not be equal, if it had remained in its expanded state. To produce continuous motion, or the up and down stroke of a piston, we must, as I have already stated, heat and cool, just as with a magnetic machine we must magnetise and demagnetise in order to produce a continuous mechanical effect; and although, from the impossibility of insulating heat, some heat is apparently lost in the process, the result may be said to be effected by the transfer of heat from the hot to the cold body, from the furnace to the condenser. But we may equally well say that the heat has been converted into mechanical force, and the mechanical force back into heat; the effects are always correlative, as are the mechanical effects of an air pump, with which, as we dilate the air on one side, we condense it on the other; and as we cannot dilate without the reciprocal condensation, so we cannot heat without the reciprocal cooling, or vice versâ.

Hitherto the resistances of the piston or of any superimposed weight have been thrown out of consideration, or, what amounts to the same thing, it has been assumed that the weight raised by the piston has descended with it. The heat has not merely been employed in dilating the air or vapour, but in raising the piston with its weight. If, as the vapour is cooled, the weight be permitted to descend, its mechanical force restores the heat lost by the dilatation; but in this case no part of the power can be abstracted so as to be employed for any practical purpose; this question then follows, what takes place with regard to the initial heat, if, after the ascent of the piston, the weight be removed so as not to help the piston in its descent, but to fall upon a lever or produce some extraneous mechanical effect?

To answer this question, let us suppose a weight to rest on a piston which confines air at a definite temperature, say for example 50°, in a cylinder, the whole being assumed to be absolutely non-conducting for heat. A part of the heat of this confined air will be due to the pressure, since, as we have seen, compression of an elastic fluid produces heat.

Suppose, now, the confined air to be heated to 70°, the piston with its superincumbent weight will ascend, and the temperature, in consequence of the dilatation of the air, will be somewhat lowered, say to 69° (we will assume, for the sake of simplicity, that the heat engendered by the friction of the piston compensates the force lost by friction).

The piston having reached its maximum of elevation, let a cold body or condenser take away 20° from the temperature of the confined air; the piston will now descend, and by the compression which the weight on it produces, will restore the 1° lost by dilatation, and when the piston reaches its original position the temperature of the air will be restored to 50°. Suppose this experiment repeated up to the rise of the piston; but when the piston is at its full elevation, and the cold body applied, let the weight be removed, so as to drop upon a wheel, or to be used for other mechanical purposes. The descending piston will not now reach its original point without more heat being abstracted; in consequence of the removal of the weight, there will not be the same force to restore the 1°, and the temperature will be 49°, or some fraction short of the original 50°. If this were otherwise, then, as the weight in falling may be made to produce heat by friction, we should have more heat than at first, or a creation of heat out of nothing—in other words, perpetual motion.

Let us now assume that this 20° supplied in the first instance was yielded by a body at 90°, of such size and material that its total capacity for heat is equal to that of the mass of confined air: this body would be reduced in temperature to 70°, in other words, our furnace would have lost 20° of heat. Let the cold body of the same size and material, used as a condenser, be at 30°. In the first experiment, the body at 30° would bring back the piston to its original point; but in the second experiment, or that where the weight has been removed, the body at 30° would not suffice to restore the piston: to effect this, the cold body or condenser must be at a lower temperature.

The question in Carnot's theory, which is not experimentally resolved, and which presents extreme experimental difficulty, is the following: Granted that a piston with a superimposed weight be raised by the thermic expansion of confined gas or vapour below it; if the elastic medium be restored to its original temperature by cooling, the weight in depressing the piston will restore that portion of the heat which has been lost by the expansion, and by the mechanical effect consequent thereon; but if the weight be removed when at its maximum of elevation, and the piston be brought back to its starting point by a necessarily cooler body than could restore it if the weight were not removed, would the return of the piston now restore the heat which had been lost by the dilatation, or, in other words, would pulling the piston down by cold restore the heat equally with the pressing it down by mechanical force? The argument from the impossibility of perpetual motion would say no, for if all the heat were restored, the mechanical effect produced by the fall of the weight, or the heating effect which might be made to result from this mechanical power, would be got from nothing.

Then follows another question, viz. whether, where an external or derived mechanical effect has been obtained, would the return of the piston, effected without the weight or external force to assist it, but solely by the colder body, give to this latter the same number of thermometric degrees as had been lost by the hot body in the first instance? Suppose, for instance, the cold body in our experiment to be at 20° instead of 30°, would this body gain 20°, and then reach the temperature of 40° when the piston is brought back, or would its temperature be higher or lower than 40°? The argument from the impossibility of perpetual motion does not apply here, for it does not necessarily follow that 20°, on the thermometric scale from 20° to 40°, represents an equal amount of force to 20° on the scale from 70° to 90°, and therefore it is quite conceivable that we may lose 20° from the furnace, and gain 20° in the condenser, and yet have obtained a certain amount of derived mechanical power. It will also follow, upon a consideration of the above imaginary experiments, that the greater the mechanical power required, the greater should be the difference between the temperature of the furnace and that of the condenser; but the exact relation in temperature between these for a given mechanical effect, has not, as far as I am aware, been satisfactorily established by experiment, though it has been shown that steam at high pressure produces, comparatively, a greater mechanical effect for the same number of degrees than steam at low pressure.

Carnot, assuming the number of degrees of temperature to be restored, but at a lower point of the thermometric scale, termed this the fall (*chute*) of caloric. The mechanical effect of heat, on this view, may be likened to that of a series of cascades on water-wheels. The highest cascade turns a wheel, and produces a given mechanical effect; the water which has produced this cannot again effect it at the same level without being carried back to its original elevation, i.e. without an extra force being employed equivalent to, or rather a fraction more than the force of the descending water; but though its power is spent with reference to the first wheel, the same water may, by falling over a new precipice upon a second wheel, again reproduce the same mechanical effect (strictly speaking, rather more, for it has approximated the centre of gravity), and so on, until no lower fall can be attained. So with heat: it involves no necessity of assuming perpetual motion to suppose that, after a given mechanical effect, produced by a certain loss of heat, the number of degrees lost from the original temperature may be restored to the condenser, but at a lower point of the thermometric scale.

If work has been done, i.e. if force has been parted with, the original temperature itself cannot be restored, but there is no *à priori* impossibility in the same number of degrees of heat as have been converted into work being conveyed to a condensing body so cold that, when it receives this heat, it will still be below the original temperature to which the work-producing heat was added.

In the theory of the steam-engine, this subject possesses a great practical interest. Watt supposed that a given weight of water required the same quantity of what is termed total heat (that is, the sensible added to the latent heat) to keep it in the state of vapour, whatever was the pressure to which it was subjected, and, consequently, however its expansive force varied. Clement Desormes was also supposed to have experimentally verified this law. If this were so, vapour raising a piston with a weight attached would produce mechanical power; and yet, the same heat existing as at first, there would be no expenditure of the initial force; and if we suppose that the heat in the condenser was the real representative of the original heat, we should get perpetual motion. Southern supposed that the latent heat was constant, and that the heat of vapour under pressure increased as the sensible heat. M. Despretz, in 1832, made some experiments, which led him to the conclusion that the increase was not in the same ratio as the sensible heat, but that yet there was an increase; a result confirmed and verified with great accuracy by M. Regnault, in some recent and elaborate researches. What seems to have occasioned the error in Watt and Clement Desormes' experiments was, the idea involved in the term latent heat; by which, supposing the phenomenon of the disappearance of sensible heat to be due to the absorption of a material substance, that substance, 'caloric,' was thought to be restored when the vapour was condensed by water, even though the water was not subjected to pressure; but to estimate the total heat of vapour under pressure the vapour should be condensed while subjected to the same pressure as that under which it is generated, as was done in M. Despretz and M. Regnault's experiments.

M. Seguin, in 1839, controverted the position that derived power could be got by the mere transfer of heat, and by calculation from certain known data, such as the law of Mariotte, viz. that the elastic force of gases and vapours increased directly with the pressure; and assuming that for vapour between 100° and 150° centigrade, each degree of elevation of temperature was produced by a thermal unit, he deduced the equivalent of mechanical work capable of being performed by a given decrement of heat; and thus concluded that, for ordinary pressures, about one gramme of water losing one degree centigrade would produce a force capable of raising a weight of 500 grammes through a space of one mètre; this estimate is a little beyond that given by the converse experiments of Mr. Joule, already stated, in which the heat produced by a given amount of mechanical action is estimated. I am not aware that the amount of mechanical work which is produced by a given quantity of heat has been directly established by experiment, though some approximative results in particular cases have been given. Theoretically it should be the same—that is to say, if a fall of 772 lbs. through a space of one foot will raise the temperature of 1 lb. of water through one degree of Fahrenheit, then the fall in the temperature of 1 lb. of water through one degree of Fahrenheit should be able to raise 772 lbs. through a space of one foot. The calculations of M. Seguin are not far from this, but since the elaborate experiments of M. Regnault he has expressed some doubt of the correctness of his former estimate, as by these experiments it appears that, within certain limits, for elevating the temperature of compressed vapour by one degree, no more than about three-tenths of a degree of total heat is required; consequently, the equivalent multiplied in this ratio would be 1,666 grammes instead of 500. Other investigators have given numbers more or less discordant; so that, without giving any opinion on their different results, this question may be considered at present far from settled. M. Regnault himself does not give the law by which the ratio of heat varies with reference to the pressure, and is still believed to be engaged in researches on the subject—one involving questions of which experiments on the mechanical effects of elastic fluids seem to offer the most promising means of solution.

I have endeavoured to give a proof (by showing the anomaly to which the contrary conclusion would lead) that, whatever amount of mechanical power is produced by one mode of application of heat, the same should, in theory, be equally produced by any other mode. But in practice the difference is immense; and therefore it becomes a question of great interest practically to ascertain what is the most convenient medium on which to apply the heat employed, and the best machinery for economising it. One great problem to be solved is the saving of the heat which the steam in ordinary engines, after having done its work, carries into the condenser, or, in the high-pressure engine, into the air. It is argued you have a large amount of fuel consumed to raise water to the boiling point, at which its efficiency as a motive agent commences. After it has done a small portion of work, and while it still retains a very large portion of the heat originally communicated to it, you reject it, and have to start again with a fresh portion of steam which has similarly exhausted fuel—in other words, you throw away all, and more than all the heat which has been employed in raising the water to the boiling point. Various plans have been devised to remedy this. Using again the warm water of the condenser to feed the boiler regains a part, but a very small part, of the heat. Employing the steam first for a high pressure, and then before its rejection or condensation using it for a low pressure, cylinder, is a second mode; a third is to use the steam, after it has done its work on the piston, as a source of heat or second furnace, to boil ether, or some liquid which evaporates at a lower temperature than water. These plans have certain advantages; but the complexity of apparatus, the danger from combustion of ether, and other reasons, have hitherto precluded their general adoption. Under the term regenerating engine various ingenious combinations have lately been suggested, and some experimental engines tried, with what success it is perhaps too early at present to pronounce an opinion. The fundamental notion on which this class of engine is based is that the vapour or air, when it has performed a certain amount of work, as by raising a piston, should, instead of being condensed or blown off, be retained and again heated to its original high temperature, and then used *de novo*; or that it should impart its heat to some other substance, and the latter in turn impart it to the fresh vapour about to act. The latter plan has been proposed by Mr. Ericsson: he passes the air which has done its work through layers of wire gauze, which are heated by the rejected air, and through which the next charge of air is made to pass. M. Seguin and Mr. Siemens have constructed machines upon the former principle, which are said to have given good experimental results. There is, however, a theoretical difficulty in all these, not affecting their capability of acting, but affecting the question of economy, which it does not seem easy to escape from. Whether the heated air or vapour be retained, or whether it yield its heat to a metallic or other substance, this heat must exercise its usual repulsive force, and this must re-act either against the returning piston or against the incoming vapour, and require a greater pressure in that to neutralise it. Vapour raising a piston and producing mechanical force effects this with decreasing power in proportion as the piston is moved. At a certain point the piston is arrested, or the stroke, as it is termed, is completed, but there is still compressed vapour in the cylinder capable of doing work, but so little that it is, and must in practice be neglected; if this compressed vapour be retained, the piston cannot be depressed without an extra force capable of overcoming the resistance of this, so to speak, semi-compressed vapour, in addition to that which is requisite to produce the normal work of the machine; and in whatever way the residual force be retained, it must either be antagonised at a loss of power for the initial force, or at most can only yield the more feeble power which it would have originally given if it had been allowed to act for a longer stroke on the piston. It may be that a portion of this residual force may be economised; indeed, this is done when the boiler is charged with warm water from the condenser, instead of with cold water; but some, indeed a notable loss, seems inevitable.

Without farther discussing the various inventions and theories on this subject, which are daily receiving increased development, it may be well to point out how far nature distances art in its present state. According to some careful estimates, the most economical of our furnaces consume from ten to twenty times as much fuel to produce the same quantity of heat as an animal produces; and Matteucci found that, from a given consumption of zinc in a voltaic battery, a far greater mechanical effect could be produced by making it act on the limbs of a recently-killed frog, notwithstanding the manifold defects of such an arrangement and its inferiority to the action of the living animal, than when the same battery was made to produce mechanical power, by acting on an electro-magnetic or other artificial motor apparatus. The ratio in his experiments was nearly six to one. Thus in all our artificial combinations we can but apply natural forces, and with far inferior mechanism to that which is perceptible in the economy of nature.

> Nature is made better by no mean;
> But nature makes that mean; so o'er that art,
> Which you say adds to nature, is an art
> That nature makes.

A speculation has been thrown out by Mr. Thompson, that, as a certain amount of heat results from mechanical action, chemical action, &c., and this heat is radiated into space, there must be a gradual diminution of temperature for the earth, by which expenditure, however slow, being continuous, it would ultimately be cooled to a degree incompatible with the existence of animal and vegetable life—in short, that the earth and the planets of our system are parting with more heat than they receive, and are therefore progressively cooling. Geological researches seemed at first to support this view, as they showed that the climate of many portions of the terrestrial surface was at remote periods hotter than at the present time: the animals whose fossilised remains are found in ancient strata have their organism adapted to what we should now term a hot climate.* There are, however, so many circumstances of difficulty attending cosmical speculations, that but little reliance can be placed upon the most profound. We know not the original source of terrestrial heat, still less that of solar heat; we know not whether or not systems of planets may be so constituted as to communicate forces, *inter se*, so that forces which have hitherto escaped detection may be in a continuous or recurring state of interchange.

The movements produced by mutual gravitation

* See *post*, p. 315.

may be the means of calling into existence molecular forces within the substances of the planets themselves. As neither from observation, nor from deduction, can we fix or conjecture any boundary to the universe of stellar orbs, as each advance in telescopic power gives us a new shell, so to speak, of stars, we may regard our globe, in the limit, as surrounded by a sphere of matter radiating heat, light, and possibly other forces.

Such stellar radiations would not, from the evidence we have at present, appear sufficient to supply the loss of heat by terrestrial radiations; but it is quite conceivable that the whole solar system may pass through portions of space having different temperatures, as was suggested, I believe, by Poisson; that as we have a terrestrial summer and winter, so there may be a solar or systematic summer and winter, in which case the heat lost during the latter period might be restored during the former. The amount of the radiations of the celestial bodies may again, from changes in their positions, vary through epochs which are of enormous duration as regards the existence of the human species.

The views of Mr. Thompson differ from those of Laplace, recently enforced by M. Babinet, which suppose the planets to have been formed by a gradual condensation of nebulous matter. A modification of this view might, perhaps, be suggested, viz. that worlds or systems, instead of being created as wholes at definite periods, are gradually changing by atmospheric additions or subtractions, or by accretions or diminutions arising from nebulous substance or from meteoric bodies, so that no star or planet could at any time be said to be created or destroyed, or to be in a state of absolute stability, but that some may be increasing, others dwindling away, and so throughout the universe, in the past as in the future. When, however, questions relating to cosmogony, or to the beginning or end of worlds, are contemplated from a physical point of view, the period of time over which our experience, in its most enlarged sense, extends, is so indefinitely minute with reference to that which must be required for any notable change, even in our own planet, that a variety of theories may be framed equally incapable of proof or of disproof. We have no means of ascertaining whether many changes, which endure in the same direction for a term beyond the range of human experience, are really continuous or only secular variations, which may be compensated for at periods far beyond our ken, so that in such cases the question of comparative stability or change can at best be only answered as to a term which, though enormous with reference to our computations, sinks into nothing with reference to cosmical time, if cosmical time be not eternity. Subjects such as these, though of a kind on which the mind delights to speculate, appear, with reference to any hope of attaining reliable knowledge, far beyond the reach of any present or immediately prospective capacity of man.

ELECTRICITY.

Electricity is that affection of matter or mode of force which most distinctly and beautifully brings into relation other modes of force, and exhibits, to a great extent in a quantitative form, its own relation with them, and their reciprocal relations with it and with each other. From the manner in which the peculiar force called electricity is seemingly transmitted through certain bodies, such as metallic wires, the term *current* is commonly used to denote its apparent progress. It is very difficult to present to the mind any theory which will give a definite conception of its *modus agendi*: the early theories regard its phenomena as produced either by a single fluid idio-repulsive but attractive of all matter, or else as produced by two fluids, each idio-repulsive but attractive of the other. No substantive theory has been proposed other than these two; but although this is the case, I think I shall not be unsupported by many who have attentively studied electrical phenomena, in viewing them as resulting, not from the action of a fluid or fluids, but as a molecular polarisation of ordinary matter, or as matter acting by attraction and repulsion in a definite direction. Thus, the transmission of the voltaic current in liquids is viewed by Grotthus as a series of chemical affinities acting in a definite direction: for instance, in the electrolysis of water, i.e. its decomposition when placed between the poles or electrodes of a voltaic battery, a molecule of oxygen is supposed to be displaced by the exalted attraction of the neighbouring electrode; the hydrogen liberated by this displacement unites with the oxygen of the contiguous molecule of water; this in turn liberates its hydrogen, and so on; the current being nothing else than this molecular transmission of chemical affinity.

There is strong reason for believing that, with some exceptions, such as fused metals, liquids do not conduct electricity without undergoing decomposition; for even in those extreme cases where a trifling effect of conduction is apparently produced without the usual elimination of substances at the electrodes, the latter when detached from the circuit show, by the counter-current which they are capable of producing when immersed in a fresh liquid, that their superficial state has been changed, doubtless by the determination to the surfaces of minute layers of substances having opposite chemical characters. The question whether or not a minute conduction in liquids can take place unaccompanied by chemical action, has however been much agitated, and may be regarded as *inter apices* of the science.

Assuming for the moment electrolysis to be the only known electrical phenomenon, electricity would appear to consist in transmitted chemical action. All the evidence we have is, that a certain affection of matter or chemical change takes place at certain distant points of space, the change at one point having a definite relation to the change at the other, and being capable of manifestation at any intermediate points.

If, now, the electrical effect called induction be examined, the phenomena will be found equally opposed to the theory of a fluid, and consistent with that of molecular polarisation. When an electrified conductor is brought near another which is not electrified, the latter becomes electrified by influence or induction, as it is termed, the nearest parts of each of these two bodies exhibiting states of electricity of the contrary denominations. Until this subject was investigated by Faraday, the intervening non-conducting body or dielectric was supposed to be purely negative, and the effect was attributed to the repulsion at a distance of the electrical fluid. Faraday showed that these effects differed greatly according to the dielectric that was interposed. Thus they were more exalted with sulphur than with shellac; more with shellac than with glass, &c. Matteucci, though differing from Faraday as to the explanation he gave, added some experiments which prove that the intervening dielectic is molecularly polarised. Thus a number of thin plates of mica are superposed like a pack of cards; metallic plates are applied to the outer facings, and one of them electrified, so that the apparatus is charged like a Leyden phial. Upon separating the plates with insulating handles, each plate is separately electrified, one side of it being positive and the other negative, showing very neatly and decisively a polarisation throughout the intervening substances by the effect of induction.

Indeed, chemical action or electrolysis may, as I have shown, be transmitted by induction across a dielectric substance, such as glass, but apparently only while the glass is being charged with electricity. A wire passing through and hermetically sealed into a glass tube, a short portion only projecting, is made to dip into water contained in a Florence flask; the flask is immersed in water to an equal depth with that within it; the wire and another similar wire dipping into the outer water are made to communicate metallically with the powerful electrical machine known as Rhumkorf's coil; bubbles of gas instantly ascend from the exposed portions of the wires, but cease after a certain time, and are renewed when, after an interval of separation, the coil is again connected with the wires.

The following interesting experiment by Mr. Karsten goes a step farther in corroboration of the molecular changes consequent upon electrisation: A coin is placed on a pack of thin plates of glass, and then electrified. On removing the coin and breathing on the glass plate, an impression of the coin is perceptible; this shows a certain molecular change on the surface of the glass opposed to the plate, or of the vapours condensed on such surface. This effect might, and has been interpreted as arising from a film of greasy deposit, supposed to exist on the plate; the impressions, however, have been proved to penetrate to certain depths below the surface, and not to be removed by polishing.

The following experiment, however, goes farther: On separating carefully the glass plates, images of the coin can be developed on each of the surfaces, showing that the molecular change has been transmitted through the substance of the glass; and we may thence reasonably suppose that a piece of glass, or other dielectric body, if it could be split up while under the influence of electric induction, would exhibit some molecular change at each side of each lamina, however minutely subdivided. I have succeeded in farther extending this experiment, and in permanently fixing the images thus produced by electricity. Between two carefully-cleaned glass plates is placed a word or device cut out of paper or tinfoil; sheets of tinfoil a little smaller than the glass plates are placed on the outside of each plate, and these coatings are brought into contact with the terminals of Rhumkorf's coil. After electrisation for a few seconds, the glasses are separated, and their interior surfaces exposed to the vapour of hydrofluoric acid, which acts chemically on glass; the portions of the glass not protected by the paper device are corroded, while those so protected are untouched or less affected by the acid, so that a permanent etching is thus produced, which nothing but disintegration of the glass will efface.

Some further experiments of mine on this subject bring out in a still more striking manner these curious molecular changes. One of the plates of glass having been electrified in the manner just mentioned, is coated, on the side impressed with the invisible electrical image, with a film of iodised collodion in the manner usually adopted for photographic purposes; it is then in a dark room immersed in a solution of nitrate of silver; then exposed to diffuse light for a few seconds. On pouring over the collodion the usual solution of pyrogallic acid, the invisible electrical image is brought out as a dark device on a light ground, and can be permanently fixed by hyposulphite of soda. The point worthy of observation in this experiment is, that this permanent image exists in the collodion film, which can be stripped off the glass, dried, and placed on any other surface, so that the molecular change consequent on electrisation has communicated, by contact or close proximity, a change to the film of collodion corresponding in form with that on the glass, but undoubtedly of a chemical nature. Electricity has, moreover, in this experiment so modified the surface of glass, that it can, in its turn, modify the structure of another substance so as to alter the relation of the latter to light. It would require a curious complication of hypothetic fluids to explain this; but if electricity and light be supposed to be affections of ordinary ponderable matter, the difficulty is only one of detail.

If, again, we examine the electricity of the atmosphere, when, as is usually the case, it is positive with respect to that of the earth, we find that each successive stratum is positive to those below it and negative to those above it; and the converse is the case when the electricity of the atmosphere is negative with respect to that of the earth.

If another electrical phenomenon be selected, another sort of change will be found to have taken place. The electric spark, the brush, and similar phenomena, the old theories regarded as actual emanations of the matter or fluid, Electricity; I venture to regard them as produced by an emission of the material itself from whence they issue, and a molecular action of the gas, or intermedium, through or across which they are transmitted.

The colour of the electric spark, or of the voltaic arc (i.e. the flame which plays between the terminal points of a powerful voltaic battery), is dependent upon the substance of the metal, subject to certain modifications of the intermedium: thus, the electric spark or arc from zinc is blue; from silver, green; from iron, red and scintillating; precisely the colours afforded by these metals in their ordinary combustion. A portion of the metal is also found to be actually transmitted with every electric or voltaic discharge: in the latter case, indeed, where the quantity of matter acted upon is greater than in the former, the metallic particles emitted by the electrodes or terminals can be readily collected, tested, or even weighed. It would thus appear that the electrical discharge arises, at least in part, from an actual repulsion and severance of the electrified matter itself, which flies off at the points of least resistance.

A careful examination of the phenomena attending the electric spark or the voltaic arc, which latter is the electric disruptive discharge acting on greater portions of matter, tends to modify considerably our previous idea of the nature of the electric force as a producer of ignition and combustion. The voltaic arc is perhaps, strictly speaking, neither ignition nor combustion. It is not simply ignition; because the matter of the terminals is not merely brought to a state of incandescence, but is physically separated and partially transferred from one electrode to another, much of it being dissipated in a vaporous state. It is not combustion; for the phenomena will take place independently of atmospheric air, oxygen gas, or any of the bodies usually called supporters of combustion, combustion being in fact chemical union attended with heat and light. In the voltaic arc we may have no chemical union; for if the experiment be performed in an exhausted receiver, or in nitrogen, the substance forming the electrodes is condensed, and precipitated upon the interior of the vessel in, chemically speaking, an unaltered state. Thus, to take a very striking example, if the voltaic discharge be taken between zinc terminals in an exhausted receiver, a fine black powder of zinc is deposited on the sides of the receiver; this can be collected, and takes fire readily in the air by being touched with a match, or ignited wire, instantly burning into white oxide of zinc. To an ordinary observer, the zinc would appear to be burned twice—first in the receiver, where the phenomenon presents all the appearance of combustion, and secondly in the real combustion in air. With iron the experiment is equally instructive. Iron is volatilised by the voltaic arc in nitrogen or in an exhausted receiver; and when a scarcely perceptible film has lined the receiver, this is washed with an acid, which then gives, with ferrocyanide of potassium, the prussian-blue precipitate. In this case we readily distil iron, a metal by ordinary means *fusible* only at a very high temperature.

Another strong evidence that the voltaic discharge consists of the material itself of which the terminals are composed, is the peculiar rotation which is observed in the light when iron is employed, the magnetic character of this metal causing its molecules to rotate by the influence of the voltaic current.

If we increase the number of reduplications in a voltaic series, we increase the length of the arc, and also increase its intensity or power of overcoming resistance. With a battery consisting of a limited number, say 100 reduplications, the discharge will not pass from one terminal to the other without first bringing them into contact, but if we increase the number of cells to 400 or 500, the discharge will pass from one terminal to the other before they are brought into contact. The difference between what is called Franklinic electricity, or that produced by an ordinary electrical machine, and voltaic electricity, or that produced by the ordinary voltaic battery, is that the former is of much greater intensity than the latter, or has a greater power of overcoming resistance, but, assuming an equal initial power, it acts upon a much smaller quantity of matter. If, then, a voltaic battery be formed with a view to increase the intensity and lessen the quantity, the character of the electrical phenomena approximate those of the electrical machine. In order to effect this, the sizes of the plates of the battery, and thence the quantity of matter acted on in each cell, must be reduced, but the number of reduplications increased. Thus if in a battery of 100 pairs of plates each plate be divided, and the battery be arranged so as to form 200 pairs, each being half the original size, the quantitative effects are diminished, and the effects of intensity increased. By carrying on this sub-division, diminishing the sizes and increasing the number, as is the case in the voltaic piles of Deluc and Zamboni, effects are ultimately produced similar to those of Franklinic electricity, and we thus gradually pass from the voltaic arc to the spark or electric discharge.

This discharge, as I have already stated, has a colour depending in part upon the nature of the terminals employed. If these terminals be highly polished, a spot will be observed, even in the case of a small electric spark, at the points from which the discharge emanates. The matter of the terminals is itself affected; and a transmission of this matter across the intervening space is detected by the deposition of minute quantities of the metal or substance composing the one, upon the other terminal.

If the gas or elastic medium between the terminals be changed, a change takes place in the length or colour of the discharge, showing an affection of the intervening matter. If the gas be rarefied, the discharge gradually changes with the degree of rarefaction, from a spark to a luminous glow or diffuse light, differing in colour in different gases, and capable of extending to a much greater distance than when it takes place in air of the ordinary density. Thus, in highly-attenuated air a discharge may be made to pass across six or seven feet of space, while in air of the ordinary density it would not pass across an inch. An observer regarding the beautiful phenomena exhibited by this electric discharge in attenuated gas, which, from some degree of similarity in appearance to the Aurora Borealis, has been called the electric Aurora, would have some difficulty in believing such effects could be due to an action of ordinary matter. The amount of gas present is extremely small; and the terminals, to a cursory examination, show no change after long experimenting. It is therefore not to be wondered at that the first observers of this and similar phenomena, regarded electricity as in itself something—as a specific existence or fluid. Even in this extreme case, however, upon a more careful examination we shall find that a change does take place, both as regards the gas and as regards the terminals. Let one of these consist of a highly-polished metal—a silver plate is one of the best materials for the purpose—

and let the discharges in attenuated atmospheric air take place from a point, say a common sewing needle, to the surface of the polished silver plate; it will be found that this is gradually changed in appearance opposite the point—it is oxidated, and gradually more and more corroded as the discharge is continued.

If now the gas be changed, and highly-rarefied hydrogen be substituted for the rarefied air, all other things remaining the same, upon passing the discharges as before the oxide will be cleared off the plate, and the polish to a great extent restored—not entirely, because the silver has been disintegrated by the oxidation—and the portion which has been affected by the discharge will present a somewhat different appearance from the remainder of the plate.

A question will probably here occur to the reader:—What will be the effect if there be not an oxidating medium present, and the experiment be first performed in a rarefied gas, which possesses no power of chemically acting on the plate? In this case there will still be a molecular change or disintegration of the plate; the portion of it acted on by the discharge will present a different appearance from that which is beyond its reach, and a whitish film, somewhat similar to that seen on the mercurialised portions of a daguerreotype, will gradually appear on the portion of the plate affected by the discharge. If the gas be a compound, as carbonic oxide, or a mixture, as oxygen and hydrogen, and consequently contain elements capable of producing oxidation and reduction, then the effect upon the plate will depend upon whether it be positive or negative; in the former case it will be oxidated, in the latter the oxide, if existing, will be reduced. This effect will also take place in atmospheric air, if it be highly rarefied, and can hardly be explained otherwise than by a molecular polarisation of the compound gas. If, again, the metal be reduced to a small point, and be of such material that the gas cannot act chemically upon it, it can yet be shown to be disintegrated by the electric spark. Thus, let a fine platinum wire be hermetically sealed in a .glass tube, and the extremity of the tube and the wire ground to a flat surface, so as to expose a section only of the wire; after taking the discharge from this for some time, it will be found that the platinum wire is worn away, and that its termination is sensibly below the level of the glass. If the discharges from such a platinum wire be taken in gas contained in a narrow tube, a cloud or film consisting of a deposit of platinum will be seen on the part of the tube surrounding the point.

Another curious effect which, in addition to the above, I have detected in the electrical discharge in attenuated media, is that when passing between terminals of a certain form, as from a wire placed at right angles to a polished plate, the discharge possesses certain phases or fits of an alternate character, so that, instead of impressing an uniform mark on a polished plate, a series of concentric rings is formed.

Priestley observed that, after the discharge of a Leyden battery, rings consisting of fused globules of metal were formed on the terminal plates; in my experiments made in attenuated media, alternate rings of oxidation and deoxidation are formed. Thus, if the plate be polished, coloured rings of oxide will alternate with rings of polished or unoxidated surface; and if the plate be previously coated with an uniform film of oxide, the oxide will be removed in concentric spaces, and increased in the alternate ones, showing a lateral alternation of positive and negative electricity, or electricity of opposite character in the same discharge.

It would be hasty to assert that in no case can the electrical disruptive discharge take place without the terminals being affected. I have, however, seen no instance of such a result where the discharge has been sufficiently prolonged, and the terminals in such a state as could be expected to render manifest slight changes.

The next question which would occur in following out the enquiry which has been indicated, would probably be, What is the action upon the gas itself? is this changed in any manner?

In answer to this, it must be admitted that, in the present state of experimental knowledge on this subject, certain gases only appear to leave permanent traces of their having been changed by the discharge, while others, if affected by it, which, as will be presently seen, there are reasons to believe they are, return to their normal state immediately after the discharge.

In the former class we may place many compound gases, as ammonia, olefiant gas, protoxide of nitrogen, deutoxide of nitrogen, and others, which are decomposed by the action of the discharge. Mixed gases are also chemically combined: for instance, oxygen and hydrogen unite and form water; common air gives nitric acid; chlorine and aqueous vapour give oxygen, the chlorine uniting with the hydrogen of the water.

But, further than this, in the case of certain elementary gases a permanent change is effected by the electrical discharge. Thus, oxygen, submitted to the discharge is partially changed into the substance called ozone, a substance now considered to be an allotropic condition of oxygen; and there is reason to believe that, when the change takes place, there is a definite polar condition of the gas, and that definite portions of it are affected—that in a certain sense one portion of the oxygen bears temporarily to the other the relation which hydrogen ordinarily does to oxygen.

If the discharge be passed through the vapour of phosphorus in the vacuum of a good air-pump, a deposit of allotropic phosphorus soon coats the interior of the receiver, showing an analogous change to that produced in oxygen; and in this case a series of transverse bands or stratifications appears in the discharge, showing a most striking alteration in its physical character, dependent on the medium across which it is transmitted. These effects were first observed by me in the year 1852. They have since been much examined by continental philosophers, and much extended by Mr. Gassiot; but no satisfactory *rationale* of them has yet been given.

There are many gases which either do not show any permanent change, or (which is more probably the case) the changes produced in them by the electrical discharge have not yet been detected. Even with these gases, however, the difference of colour, of length, or of the different position of a certain dark space or spaces which appear in the discharge, show that the discharge differs for different media. We never find that the discharge has itself added to or subtracted from the total weight of the substances acted on: we find no evidence of a fluid but the visible phenomena themselves; and those we may account for by the change which takes place in the matter affected.

I have here, as elsewhere, used words of common acceptation, such as 'matter affected by the discharge,' &c., though upon the view I am suggesting, the discharge is itself this affection of matter; and the writing these passages affords, to me at least, a striking instance of how much ideas are bound up in words, when, to express a view differing from the received one, words involving the received one are necessarily used.

Passing now to the effect of the transmission of electricity by the class of the best conducting bodies, such as the metals and carbon, here, though we cannot at present give the exact character of the motion impressed upon the particles, there are yet many experiments which show that a change takes place in such substances when they are affected by electricity.

Let discharges from a Leyden jar or battery be passed through a platinum wire, too thick to be fused by the discharges, and free from constraint, it will be found that the wire is shortened; it has undergone a molecular change, and apparently been acted on by a force transverse to its length. If the discharges be continued, it gradually gathers up in small irregular bends or convolutions. So with voltaic electricity: place a platinum wire in a trough of porcelain, so that when fused it shall retain its position as a wire, and then ignite it by a voltaic battery. As it reaches the point of fusion it will snap asunder, showing a contraction in length, and consequently a distension or increase in its transverse dimensions. Perform the same experiment with a lead wire, which can be more readily kept in a state of fusion, and follow it, as it contracts, by the terminal wires of the battery; it will be seen to gather up in nodules, which press on each other like a string of beads of a soft material which have been longitudinally compressed.

As we increase the thickness of the wires in these experiments with reference to the electrical force employed, we lessen the perceptible effect; but even in this case we shall be enabled safely to infer that some molecular change accompanies the transmission of electricity: the wires are heated in a degree decreasing as their thickness increases— but by decreasing the delicacy of our tests as the heating effects decrease in intensity, we may indefinitely detect the augmentation of temperature accompanying the passage of electricty—and wherever there is augmentation of temperature there must be expansion or change of position of the molecules.

Again, it has been observed that wires which have for a long time transmitted electricity, such as those which have served as conductors for atmospheric electricity, have their texture changed, and are rendered brittle. In this observation, however, though made by a skilful electrician, M. Peltier, the effects of exposure to the atmosphere, to changes of temperature, &c., have not been sufficiently eliminated to render it worthy of entire confidence. There are, however, other experiments which show that the elasticity of metals is changed by the passage through them of the electric current.

Thus M. Wertheim has, from an elaborate series of experiments, arrived at the conclusion that there is a temporary diminution in the coefficient of elasticity in wires while they are transmitting the electric current, which is independent of the heating effect of the current.

M. Dufour has made a considerable number of experiments with the view of ascertaining if any permanent change in metals is effected by electrisation. He arrives at the curious result that in a copper wire through which a feeble voltaic current has passed for several days, a notable diminution in tenacity takes place; while, in an iron wire, the tenacity is increased; and that these effects were more perceptible when the wires had been electrised for a long time (nineteen days) than for a short time (four days). The copper wire was, in his experiment, not perfectly pure; so that the effect, or a portion of it, might be due to the state of alloy: in the case of iron, the magnetic character of the metal would probably modify the effects, and might account for the opposite character of the results with these two metals.

Matteucci has made experiments on the conduction of electricity by bismuth in directions parallel or transverse to the planes of principal cleavage, and he finds that bismuth conducts electricity and heat better in the direction of the cleavage planes than in that transverse to them.

Many other experiments have been made both on the production of thermo-electric currents by two portions of the same crystalline metal, but with the planes of crystallization arranged in different directions relatively to each other, and also on the differences in conduction of heat and electricity according to the direction in which they are transmitted with reference to the planes of crystallization.

It is found, moreover, that the slightest difference in homogeneity in the same metal enables it when heated to produce a thermo-electric current, and that metals in a state of fusion, in which state they may be presumed to be homogeneous throughout, give no thermo-electric current: thus, hot in contact with cold mercury has been shown by Matteucci to give no thermo-electric current, and the same is the case with portions of fused bismuth unequally heated.

The fact that the molecular structure or arrangement of a body influences—indeed I may say determines—its conducting power, is by no means explained by the theory of a fluid; but if electricity be only a transmission of force or motion, the influence of the molecular state is just what would be expected. Carbon, in a transparent crystalline state, as diamond, is as perfect a non-conductor as we know; while in an opaque amorphous state, as graphite or charcoal, it is one of the best conductors: thus, in the one state, it transmits light and stops electricity, in the other it transmits electricity and stops light.

It is a circumstance worthy of remark, that the arrangement of molecules, which renders a solid body capable of transmitting light, is most unfavourable to its transmission of electricity, transparent solids being very imperfect conductors of electricity; so all gases readily transmit light, but are amongst the worst conductors of electricity, if indeed, properly speaking, they can be said to conduct at all.

The conduction of electricity by different classes of bodies has been generally regarded as a question of degree: thus metals were viewed as perfect conductors, charcoal less so, water and other liquids as imperfect conductors, &c. But, in fact, though between one metal and another the mode of transmission may be the same and the difference one of degree, a different molecular effect obtains, when we contrast metals with electrolytic liquids and these with gases.

Attenuated gases may be, in one sense, regarded as non-conductors, in another as conductors; thus if gold leaves be made to diverge, by electrical repulsion, in air at ordinary pressure, they in a short time collapse; while in highly-rarefied air, or what is commonly termed a vacuum, they remain divergent for days; and yet electricity of a certain degree of tension passes readily across attenuated air, and with difficulty across air of ordinary density.

Again, where the electrical terminals are brought to a state of visible ignition, there are symptoms of the transmission of electricity of low tension across gases; but no such effects have been detected at lower temperatures. All this presents a strong argument in favour of the transmission of electricity across gases being effected by the disruptive discharge, and not by a conduction similar to that which takes place with metals or with electrolytes.

The ordinary attractions and repulsions of electrified bodies present no more difficulty when regarded as being produced by a change in the state or relations of the matter affected, than do the attractions of the earth by the sun, or of a leaden ball by the earth; the hypothesis of a fluid is not considered necessary for the latter, and need not be so for the former class of phenomena. How the phenomena are produced to which the term attraction is applied is still a mystery. Newton, speaking of it, says, 'What I call attraction may be performed by impulse, or by some other means unknown to me. I use that word here to signify only in general any force by which bodies tend towards one another, whatsoever be the cause." If we suppose a fluid to act in attractions and repulsions, the imponderable fluid must drag or push the matter with it: thus when we feel a stream of air rushing from an electrified metallic point, each molecule of air contiguous to the point being repelled, another takes its place, which is in its turn repelled;—how does a hypothetic fluid assist us here? If we say the electrical fluid repels itself, or the same electricity repels itself, we must go farther and assert, that it not only repels itself, but either communicates its repulsive force to the particles of the air, or carries with it the particle of air in its passage. Is it not more easy to assume that the particle of air is in such a state that the ordinary forces which keep it in equilibrium are disturbed by the electrical force, or force in a definite direction communicated to it, and that thus each particle in turn recedes from the point? As this latter force is increased, not only does the particle of air which was contiguous to the metallic point recede, but the cohesion of the extreme particles of metal may be overcome to such an extent that these are detached, and the brush or spark may consist wholly or in part of minute particles of the metal itself thrown off. Of this there is some evidence, though the point can hardly be considered as proved. A similar effect undoubtedly

K

takes place with voltaic electricity, acting upon a terminal immersed in a liquid; thus if metallic terminals of a powerful voltaic battery be immersed in water, metal, or the oxide of metal, is forcibly detached, producing great heat at the point of disruption.

If we apply ourselves to the effect of electricity in the animal economy, we find that the first rationale given of the convulsive effect produced by transmission through the living or recently killed animal was, that electricity itself, something substantive, passed rapidly through the body, and gave rise to the contractions; step by step we are now arriving at the conviction that consecutive particles of the nerves and muscles are affected. Thus the contractions which the prepared leg of a frog undergoes at the moment it is submitted to a voltaic current, cease after a time if the current be continued, and are renewed on breaking the circuit, i. e. at the moment when the current ceases to traverse it. The excitability of a nerve, moreover, or its power of producing muscular contraction, is weakened or destroyed by the transmission of electricity in one direction, while the excitability is increased by the transmission of electricity in the opposite direction; showing that the fibre or matter itself of the nerve is changed by electrisation, and changed in a manner bearing a direct relation to the other effects produced by electricity.

Portions of muscle and of nerve present different electrical states with reference to other portions of the same muscle or nerve; thus the external part of a muscle bears the same relation to the internal part as platinum does to zinc in the voltaic battery; and delicate galvanoscopes will show electrical effects when interposed in a conducting circuit connecting the surface of a nerve with its interior portions. Matteucci has proved that a species of voltaic pile may be formed by a series of slices of muscle, so arranged that the external part of one slice may touch the internal part of the next, and so on.

Lastly, the magnetic effects produced by electricity also show a change in the molecular state of the magnetic substance affected; as we shall see when the subject of magnetism is discussed.

I have taken in succession all the known classes of electrical phenomena; and, as far as I am aware, there is not a single electrical effect, where, if a close investigation be instituted, and the materials chosen in a state for exhibiting minute changes, evidence of molecular change will not be detected; so that, excepting those cases where infinitesimally small quantities of matter are acted on, and our means of detection fail, electrical effects are known to us only as changes of ordinary matter. It seems to me as easy to imagine these changes to be effected by a force acting in definite directions, as by a fluid which has no independent or sensible existence, and which, it must be assumed, is associated with, or exerts a force acting upon ordinary matter, or matter of a different order from the supposed fluid. As the idea of the hypothetic fluid is pursued, it gradually vanishes, and resolves itself into the idea of force. The hypothesis of matter without weight presents in itself, as I believe, fatal objections to the theories of electrical *fluids*, which are entirely removed by viewing electricity as force, and not as matter.

If it be said that the effects we have been considering may still be produced by a fluid, and that this fluid acts upon ordinary matter in certain cases, polarising the matter affected or arranging its particles in a definite direction, whilst in others, by its attractive or repulsive force, it carries with it portions of matter; then, if the fluid in itself be incapable of recognition by any test, if it be only evidenced by the changes which it operates in ponderable matter, the words fluid and force become identical in meaning; we may as well say that the attraction of gravitation or weight is occasioned by a fluid, as that electrical changes are so.

When, as is constantly done in common parlance, a house is said to be *struck*, windows *broken*, metals *fused* or *dissipated* by the electrical *fluid*, are not the expressions used such as, if not sanctioned by habit, would seem absurd? In all the cases of injury done by lightning there is no fluid perceptible; the so-called sulphurous odour is either ozone developed by the action of electricity on atmospheric air, or the vapour of some substance dissipated by the discharge; does it not then seem more consonant with experience to regard these effects as produced by force, as we have analogous effects produced by admitted forces, in cases where no one would invoke the aid of a hypothetic fluid for explanation. For instance, glasses may be broken by electrical discharges; so may they by sonorous vibrations. Metals electrified or magnetised will emit a sound; so they will if struck, or if a musical note with which they can vibrate in unison be sounded near to them.

Even chemical decomposition, in cases of feeble affinity, may be produced by purely mechanical effects. A number of instances of this have been collected by M. Becquerel; and substances whose constituents are held together by feeble affinities, such as iodide of nitrogen and similar compounds, are decomposed by the vibration occasioned by sound.

If, instead of being regarded as a fluid or imponderable matter *sui generis*, electricity be regarded as the motion of an ether, equal difficulties are encountered. Assuming ether to pervade the pores of all bodies, is the ether a conductor or nonconductor? If the latter—that is, if the ether be incapable of transmitting the electrical wave—the ethereal hypothesis of electricity necessarily fails; but if the motion of the ether constitute what we call conduction of electricity, then the more porous bodies, or those most permeable by the ether, should be the best conductors. But this is not the case. If, again, the metal and the air surrounding it are both pervaded by ether, why should the electrical wave affect the ether in the metal, and not stir that in the gas? To support an ethereal hypothesis of electricity, many additional and hardly reconcilable hypotheses must be imported.

The fracture and comminution of a non-conducting body, the fusion or dispersion of a metallic wire by the electrical discharge, are effects equally difficult to conceive upon the hypothesis of an ethereal vibration, as upon that of a fluid, but are necessary results of the sudden subversion of molecular polarisation, or of a sudden or irregular vibratory movement of the matter itself. We see similar effects produced by sonorous vibrations, which might be called conduction and non-conduction of sound. One body transmits sound easily, another stops or deadens it, as it is termed—i.e. disperses the vibrations, instead of continuing them in the same direction as the primary impulse; and solid bodies may, as has been above observed, be shivered by sudden impulses of sound in those cases where all the parts of the body cannot uniformly carry on the undulatory motion.

The progressive stages in the History of Physical Philosophy will account in a great measure for the adoption by the early electricians of the theories of fluids.

The ancients, when they witnessed a natural phenomenon, removed from ordinary analogies, and unexplained by any mechanical action known to them, referred it to a soul, a spiritual or preternatural power: thus amber and the magnet were supposed by Thales to have a soul; the functions of digestion, assimilation, &c., were supposed by Paracelsus to be effected by a spirit (the Archæus). Air and gases were also at first deemed spiritual, but subsequently became invested with a more material character; and the word gas, from *geist*, a ghost or spirit, affords us an instance of the gradual transmission of a spiritual into a physical conception.

The establishment by Torricelli of the ponderable character of air and gas, showed that substances which had been deemed spiritual and essentially different from ponderable matter were possessed of its attributes. A less superstitious mode of reasoning ensued, and now aëriform fluids were shown to be analogous in many of their actions to liquids or known fluids. A belief in the existence of other fluids, differing from air as this differed from water, grew up, and when a new phenomenon presented itself, recourse was had to a hypothetic fluid for explaining the phenomenon and connecting it with others; the mind once possessed of the idea of a fluid, soon invested it with the necessary powers and properties, and grafted upon it a luxuriant vegetation of imaginary offshoots.

In what I am here throwing out, I wish to guard myself from being supposed to state that the course of theory, historically viewed, followed exactly the dates of the discoveries which were effectual in changing its character; sometimes a discovery precedes, at other times it succeeds a change in the general course of thought; sometimes, and perhaps most frequently, it does both—i.e. the discovery is the result of a tendency of the age and of the continually improved methods of observation, and when made, it strengthens and extends the views which have led to it. I think the phases of thought which physical philosophers have gone through, will be found generally such as I have indicated, and that the gradual accumulation of discoveries which has taken place during the more recent periods, by showing what effects can be produced by dynamical causes alone, is rapidly tending to a general dynamical theory into which that of the imponderable fluids promises ultimately to merge.

Commencing with electricity as an initiating force, we get *motion* directly produced by it in various forms; for instance, in the attraction and repulsion of bodies, evidenced by mobile electrometers, such as that of Cuthbertson, where large masses are acted on; the rotation of the fly-wheel, another form of electrical repulsion, and the deflection of the galvanometer needle, are also modes of palpable, visible motion.

It would follow, from the reasoning in this essay, that when electricity performs any mechanical work which does not return to the machine, electrical power is lost. It would be unsuitable to the scope of this work to give the mathematical labours of M. Clausius and others here; but the following experiment, which I devised for making the result evident to an audience at the Royal Institution, will form a useful illustration :—A Leyden jar, of one square foot coated surface, has its interior connected with a Cuthbertson's electrometer, between which and the outer coating of the jar are a pair of discharging balls fixed at a certain distance (about half an inch apart). Between the Leyden jar and the prime conductor is inserted a small unit jar of nine inches surface, the knobs of which are 0·2 inch apart.

The balance of the electrometer is now fixed by a stiff wire inserted between the attracting knobs, and the Leyden jar charged by discharges from the unit jar. After a certain number of these, say twenty, the discharge of the large jar takes place across the half-inch interval. This may be viewed as the expression of electrical power received from the unit jar. The experiment is now repeated, the wire between the balls having been removed, and therefore the 'tip,' or the raising of the weight, is performed by the electrical repulsion and attraction of the two pairs of balls. At twenty discharges of the unit jar the balance is subverted, and one attracting knob drops upon the other; but *no discharge takes place*, showing that some electricity has been lost or converted into the mechanical power which raised the balance.

By another mode of expression, the electricity may be supposed to be masked or analogous to latent heat, and it would be restored if the ball were brought back without discharge by extraneous force. If the discharge or other electrical effects were the same in both cases, then, since the raising of the ball or weight is an extra mechanical effort, and since the weight is capable by its fall of producing electricity, heat, or other force, it would seem that force could be got out of nothing, or perpetual motion obtained.

The above experiment is suggestive of others of a similar character, which may be indefinitely varied. Thus I have found that two balls made to diverge by electricity do not give to an electrometer the same amount of electricity as they do if, whilst similarly electrified, they are kept forcibly together. This experiment is the converse of the former one. There is an advantage in electrical experiments of this class as compared with those on heat, viz. that though there is no perfect insulation for electricity, yet our means of insulating it are immeasurably superior to any attainable for heat.

Electricity directly produces *heat*, as shown in the ignited wire, the electric spark, and the voltaic arc: in the latter the most intense heat with which we are acquainted—so intense, indeed, that it cannot be measured, as every sort of matter is dissipated by it.

In the phenomenon of electrical ignition, as shown by a heated conjunctive wire, the relation of force and resistance, and the correlative character of the two forces, electricity and heat, are strikingly demonstrated. Let a thin wire of platinum join the terminals of a voltaic battery of suitable power, the wire will be ignited, and a certain amount of chemical action will take place in the cells of the battery—a definite quantity of zinc being dissolved and of hydrogen eliminated in a given time. If now the platinum wire be immersed in water, the heat will, from the circulating currents of the liquid, be more rapidly dissipated, and we shall instantly find that the chemical action in the battery will be increased, more zinc will be dissolved, and more hydrogen eliminated for the same time; the heat being conveyed away by the water, more chemical action is required to generate it, just as more fuel is required in proportion as evaporation is more rapid.

Reverse the experiment, and instead of placing the wire in water, place it in the flame of a spirit lamp, so that the force of heat meets with greater resistance to its dissipation. We now find that the chemical action is less than in the first or normal experiment. If the wire be placed in other different gaseous or liquid media, we shall find that the chemical action of the battery will be proportioned to the facility with which the heat is circulated or radiated by these media, and we thus establish an alternating reciprocity of action between these two forces: a similar reciprocity may be established between electricity and motion, magnetism and motion, and so of other forces. If it cannot be realised with all, it is probably because we have not yet eliminated interfering actions. If we carefully think over the matter, we shall, unless I am much mistaken, arrive at the conclusion that it cannot be otherwise, unless it be supposed that a force can arise from nothing—can exist without antecedent force.

In the phenomenon of the voltaic arc, the electric spark, &c., to which I have already adverted, electricity directly produces *light* of the greatest intensity of any artificially obtained. It directly produces *magnetism*, as shown by Oersted, who first distinctly proved the connection between electricity and magnetism. These two forces act upon each other, not in straight lines, as all other known forces do, but in a rectangular direction; that is, bodies affected by dynamic electricity, or the conduits of an electric current, tend to place magnets at right angles to them; and, conversely, magnets tend to place bodies conducting electricity at right angles to them. Thus an electric current appears to have a magnetic action, in a direction cutting its own at right angles; or, supposing its section to be a circle, tangential to it: if, then, we reverse the position, and make the electric current form a series of tangents to an imaginary cylinder, this cylinder should be a magnet. This is effected in practice by coiling a wire as a helix or spiral, and this, when conducting an electrical current, is to all intents and purposes a magnet. A soft iron core placed within such a helix has the property of concentrating its power, and then we can, by connection or disconnection with the source of electricity, instantly make or unmake a most powerful magnet.

We may figure to the mind electrified and magnetised matter, as lines of which the extremities repel each other in a definite direction; thus, if a line A B represent a wire affected by electricity, and superposed on C D a wire affected by magnetism, the extreme points A and B will be repelled to the farthest distances from the points C and D, and the line A B be at right angles to the line C D; and so, if the lines be subdivided to any extent, each will have two extremities or poles repulsive of those of the other. If the line of matter affected by electricity be a liquid, and consequently have entire mobility of particles, a continuous movement will be produced by magnetisation, each particle successively tending, as it were, to fly off at a tangent from the magnet: thus, place a flat dish containing acidulated water on the poles of a powerful magnet, immerse the terminals of a voltaic battery in the liquid just above the magnetic poles, so that the lines of electricity and of magnetism coincide; the water will now assume a movement at right angles to this line, flowing continuously, as if blown by an equatorial wind, which may be made east or west with reference to the magnetic poles by altering the direction of the electrical current: a similar effect may be produced with mercury. These cases afford an additional argument to those previously mentioned, of the particles of matter being affected by the forces of electricity and magnetism in a way irreconcilable with the fluid or ethereal hypothesis.

The representation of transverse direction by magnetism and electricity appears to have led Coleridge to parallel it by the transverse expansion of matter, or length and breadth, though he injured the parallel by adding galvanism as depth: whether a third force exists which may bear this relation to electricity and magnetism is a question upon which we have no evidence.

The ratio which the attractive magnetic force produced bears to the electric current producing it has been investigated by many experimentalists and mathematicians. The data are so numerous and so variable, that it is difficult to arrive at definite results. Thus the relative size of the coil and the iron, the temper or degree of hardness of the latter, its shape, or the proportion of length to diameter, the number of coils surrounding it, the conducting power of the metal of which the coils are formed, the size of the keeper or iron in which magnetism is induced, the degree of constancy of the battery, &c., complicate the experiments.

The most trustworthy general relation which has been ascertained is, that the magnetic attraction is as the square of the electric force; a result due to the researches of Lenz and Jacobi, and also of Sir W. S. Harris.

Lastly, electricity produces *chemical affinity*; and by its agency we are enabled to obtain effects of analysis or synthesis with which ordinary chemistry does not furnish us. Of these effects we have examples in the brilliant discoveries, by Davy, of the alkaline metals, and in the peculiar crystalline compounds made known by Crosse and Becquerel.

LIGHT.

In entering on the subject of LIGHT, it will be well to describe briefly, and in a manner as far as may be independent of theory, the effects to which the term polarisation has been applied.

When light is reflected from the surface of water, glass, or many other media, it undergoes a change which disables it from being again similarly reflected in a direction at right angles to that at which it has been originally reflected. Light so affected is said to be polarised; it will always be capable of being reflected in planes parallel to the plane in which it has been first reflected, but incapable of being reflected in planes at right angles to that plane. At planes having a direction intermediate between the original plane of reflection, and a plane at right angles to it, the light will be capable of being partially reflected, and more or less so according as the direction of the second plane of reflection is more or less coincident with the original plane. Light, again, when passed through a crystal of Iceland spar, is what is termed doubly refracted, i.e. split into two divisions or beams, each having half the luminosity of the original incident light; each of these beams is polarised in planes at right angles to each other; and if they be intercepted by the mineral tourmaline, one of them is absorbed, so that only one polarised beam emerges. Similar effects may be produced by certain other reflections or refractions. A ray of light once polarised in a certain plane continues so affected throughout its whole subsequent course; and at any indefinite distance from the point where it originally underwent the change, the direction of the plane will be the same, provided the media through which it is transmitted be air, water, or certain other transparent substances which need not be enumerated. If, however, the polarised ray, instead of passing through water, be made to pass through oil of turpentine, the definite direction in which it is polarised will be found to be changed; and the change of direction will be greater according to the length of the column of interposed liquid. Instead of being an uniform plane, it will have a curvilinear direction, similar to that which a strip of card would have if forced along two opposite grooves of a rifle-barrel. This curious effect is produced in different degrees by different media. The direction also varies; the rotation, as it is termed, being sometimes to the right hand and sometimes to the left, according to the peculiar molecular character of the medium through which the polarised ray is transmitted. These effects will be presently reverted to.

Light is, perhaps, that mode of force the reciprocal relations of which with the others have been the least traced out. Until the discoveries of Niepce, Daguerre, and Talbot, very little could be definitely predicated of the action of light in producing other modes of force. Certain chemical compounds, among which stand pre-eminent the salts of silver, have the property of suffering decomposition when exposed to light. If, for instance, recently formed chloride of silver be submitted to luminous rays, a partial decomposition ensues; the chlorine is separated and expelled by the action of light, and the silver is precipitated. By this decomposition the colour of the substance changes from white to blue. If now, paper be impregnated with chloride of silver, which can be done by a simple chemical process, then partially covered with an opaque substance, a leaf for example, and exposed to a strong light, the chloride will be decomposed in all those parts of the paper where the light is *not* intercepted, and we shall have, by the action of light, a white image of the leaf on a purple ground. If similar paper be placed in the focus of a lens in a camera-obscura, the objects there depicted will decompose the chloride, just in the proportion in which they are luminous;

and thus, as the most luminous parts of the image will most darken the chloride, we shall have a picture of the objects with reversed lights and shadows. The picture thus produced would not be permanent, as subsequent exposure would darken the light portion of the picture: to fix it, the paper must be immersed in a solution which has the property of dissolving chloride of silver, but not metallic silver. Iodide of potassium will effect this; and the paper being washed and dried will then preserve a permanent image of the depicted objects. This was the first and simple process of Mr. Talbot; but it is defective as to the purposes aimed at, in many points. First, it is not sufficiently sensitive, requiring a strong light and a long time to produce an image; secondly, the lights and shadows are reversed; and thirdly, the coarse structure of the finest paper does not admit of the delicate traces of objects being distinctly impressed. These defects have been to a great extent remedied by a process subsequently discovered by Mr. Talbot, and which bears his name, and which has led to the collodion process, and others unnecessary to be detailed here.

The photographs of M. Daguerre, with which all are now familiar, are produced by holding a plate of highly-polished silver over iodine. A thin film of iodide of silver is thus formed on the surface of the metal; and when these iodized plates are exposed in the camera, a chemical alteration takes place. The portions of the plate on which the light has impinged part with some of the iodine, or are otherwise changed—for the theory is somewhat doubtful—so as to be capable of ready amalgamation. When, therefore, the plate is placed over the vapour of heated mercury, the mercury attaches itself to the portions affected by light, and gives them a white frosted appearance; the intermediate tints are less affected, and those parts where no light has fallen, by retaining their original polish, appear dark: the iodide of silver is then washed off by hyposulphite of soda, which has the property of dissolving it, and there remains a picture in which the lights and shadows are as in nature, and the molecular uniformity of the metallic surface enables the most microscopic details to be depicted with perfect accuracy. By using chloride of iodine, or bromide of iodine, instead of iodine, the equilibrium of chemical forces is rendered still more unstable, so that images may be taken in an indefinitely short period—a period practically instantaneous.

It would be foreign to the object of this essay to enter upon the many beautiful details into which the science of photography has branched out, and the many valuable discoveries and practical applications to which it has led. The short statement I have given above is perhaps superfluous, as, though they were new and surprising at the period when these Lectures were delivered, photographic processes have now become familiar, not only to the cultivator of science, but to the artist and amateur: the important point for consideration here is that light will chemically or molecularly affect matter. Not only will the particular compounds above selected as instances be changed by the action of light; but a vast number of substances, both elementary and compound, are notably affected by this agent, even those apparently the most unalterable in character, such as metals: so numerous, indeed, are the substances affected, that it has been supposed, not without reason, that matter of every description is altered by exposure to light.

The permanent impression stamped on the molecules of matter by light can be made to repeat itself by the same agency, but always with decreasing force. Thus a photograph placed opposite a camera containing a sensitive plate will be reproduced, but if the size of the image be equal to the picture, the second picture will be fainter than the first, and so on. Thus again, a photograph taken on a dull day cannot, by being placed in bright sunshine, be made to reproduce a second photograph of the same size and more distinctly marked than itself; I at least have never succeeded in such reproduction, and I am not aware that others have:

the image loses in intensity as light itself does by each transmission. The surface of the metal or paper may give a brighter image from its being exposed to a more intense light, but the photographic details are limited to the intensity of the first impression, or rather to something short of this. A question of theoretical interest arises from the consideration of these reproduced photographs. We know that the luminosity of the image at the focus of a telescope is limited by the area of the object-glass. The image of any given object cannot be intensified by throwing upon it extraneous light; it is indeed diminished in intensity, and when for certain purposes astronomers illuminate the fields of their telescopes, they are obliged to be contented with a loss of intensity in the telescopic image.

Now, let us suppose that the minutest details in the image of an object seen in a given telescope, and with a given power, are noted; that then a photographic plate is placed in the focus of the same telescope so as to obtain a permanent impression of the image which has been viewed by the eye-glass. Could the observer, by throwing a beam of condensed light upon the photograph, enable himself to bring out fresh details? or in other words, could he use with advantage a higher power applied to the illuminated photograph?

It is, perhaps, hardly safe to answer *à priori* this question; but the experiment of reproducing photographs would seem to show that more than the initial light cannot be got, and that we cannot expect to increase telescopic power by photography, though we may render observations more convenient, may by its means fix images seen on rare and favourable occasions, and may preserve permanent and infallible records of the past state of astronomical objects.

The effect of light on chemical compounds affords us a striking instance of the extent to which a force, ever active, may be ignored through successive ages of philosophy. If we suppose the walls of a large room covered with photographic apparatus, the small amount of light reflected from the face of a person situated in its centre would simultaneously imprint his portrait on a multitude of recipient surfaces. Were the cameras absent, but the room coated with photographic paper, a change would equally take place in every portion of it, though not a reproduction of form and figure. As other substances not commonly called photographic are known to be affected by light, the list of which might be indefinitely extended, it becomes a curious object of contemplation to consider how far light is daily operating changes in ponderable matter—how far a force, for a long time recognised only in its visual effects, may be constantly producing changes in the earth and atmosphere, in addition to the changes it produces in organised structures which are now beginning to be extensively studied. Every portion of light may be supposed to write its own history by a change more or less permanent in ponderable matter.

The late Mr. George Stephenson had a favourite idea, which would now be recognised as more philosophical than it was in his day, viz. that the light, which we nightly obtain from coal or other fuel, was a reproduction of that which had at one time been absorbed by vegetable structures from the sun. The conviction that the transient gleam leaves its permanent impress on the world's history, also leads the mind to ponder over the many possible agencies of which we of the present day may be as ignorant as the ancients were of the chemical action of light.

I have used the term light, and affected by light, in speaking of photographic effects; but, though the phenomena derived their name from light, it has been doubted by many competent investigators whether the phenomena of photography are not mainly dependent upon a separate agent accompanying light, rather than upon light itself. It is, indeed, difficult not to believe that a picture, taken in the focus of a camera-obscura, and which represents to the eye all the gradations of light and shade shown by the original luminous image, is not an effect of light; certain it is, however, that the different coloured rays exercise different actions upon various chemical compounds, and that the effects on many, perhaps on all of them, are not proportionate in intensity to the effects upon the visual organs. Those effects, however, appear to be more of degree than of specific difference; and, without pronouncing myself positively upon the question, hitherto so little examined, I think it will be safer at present to regard the action on photographic compounds as resulting from a function of light. So viewing it, we get light as an initiating force, capable of producing, mediately or immediately, the other modes of force. Thus, it immediately produces chemical action; and having this, we at once acquire a means of producing the others. At my Lectures in 1843, I showed an experiment by which the production of all the other modes of force by light is exhibited: I may here shortly describe it. A prepared daguerreotype plate is enclosed in a box filled with water, having a glass front with a shutter over it. Between this glass and the plate is a gridiron of silver wire; the plate is connected with one extremity of a galvanometer coil, and the gridiron of wire with one extremity of a Breguet's helix—an elegant instrument, formed by a coil of two metals, the unequal expansion of which indicates slight changes in temperature— the other extremities of the galvanometer and helix are connected by a wire, and the needles brought to zero. As soon as a beam of either daylight or the oxyhydrogen light is, by raising the shutter, permitted to impinge upon the plate, the needles are deflected. Thus light being the initiating force, we get *chemical action* on the plate, *electricity* circulating through the wires, *magnetism* in the coil, *heat* in the helix, and *motion* in the needles.

If two plates of platinum be placed in acidulated water, and connected with a delicate galvanometer, the needle of this is always deflected, a result due to films of gas or other matter on the surface of the platinum, which no cleaning can remove. If, after the needle has returned to zero, which will not be the case for some hours or even days, one of the platinum surfaces be exposed to light, a fresh deflection of the needle takes place, due, as far as I have been able to resolve it, to an augmentation of the chemical action which had occasioned the orginal deflection, for the deviation is in the same direction. If, instead of white light, coloured light be permitted to impinge on the plate, the deviation is greater with blue than with red or yellow light, showing, in addition to other tests, that the effect is not due to the heat of the sun's rays, as the calorific effects of light are greater with red than with blue light, while the chemical effects are the inverse.

There are other apparently more direct agencies of light in producing electricity and magnetism, such as those observed by Morichini and others, as well as its effects upon crystallization; but these results have hitherto been of so indefinite a character, that they can only be regarded as presenting fields for experiment, and not as proving the relations of light to the other forces.

Light would seem directly to produce heat in the phenomena of what is termed absorption of light: in some of these we find that heat is developed in proportion to the disappearance of light. It was thought, and Franklin's experiment of different coloured cloths on snow exposed to the sun seemed to prove it, that heat was produced in a direct ratio to the absorption of light, but Dr. Tyndall has shown that this result does not depend on the colour, but on the chemical and physical character of the substance exposed, and that the heating effect is due mainly to non-luminous radiation. The heating powers of different colours are by no means in exact proportion to the intensity of their light as affecting the visual organs. Thus red light, when produced by refraction from a prism of glass, produces greater heating effect than yellow light in the phenomena of absorption, as has been observed by Sir W. Herschel. The red rays appear, however, to produce a dynamic effect greater than any of the others; thus they penetrate water to a greater depth than the other colours; but, according to Dr. Seebeck, we get a further anomaly, viz. that when light is refracted by a prism of water the yellow rays produce the greater heating effect. The subject, therefore, requires much more experiment before we can ascertain the rationale of the action of the forces of light and heat in this class of phenomena.

In a former edition of this Essay, I suggested the following experiment on this subject:—Let a beam of light be passed through two plates of tourmaline, or similar substance, and the temperature of the second plate, or that on which the light last impinges, be examined by a delicate thermoscope, first when it is in a position to transmit the polarised beam coming from the first plate, and secondly when it has been turned round through an arc of 90°, and the polarised beam is absorbed. I expected that, if the experiment were carefully performed, the temperature of the second plate would be more raised in the second case than in the first, and that it might afford interesting results when tried with light of different colours. I met with difficulties in procuring a suitable apparatus, and was endeavouring to overcome them when I found that Knoblauch had, to some extent, realised this result. He finds that, when a solar beam, polarised in a certain plane, is transmitted perpendicularly to the axis of a crystal of brown quartz or tourmaline, the heat is transmitted in a smaller proportion than when the beam passes along the direction of the axis of the crystal.

It is generally—as far as I am aware, universally—true that, while light continues as light, even though reflected or transmitted by different media, little or no heat is developed; and, as far as we can judge, it would appear that, if a medium were perfectly transparent, or if a surface perfectly reflected light, not the slightest heating effect would take place; but, wherever light is absorbed, then heat takes its place, affording us apparently an instance of the conversion of light into heat, and of the fact that the force of light is not, in fact, absorbed or annihilated, but merely changed in character, becoming in this instance converted into heat by impinging on solid matter, as in the instance mentioned in treating of heat, this force was shown to be converted into light by impinging on solid matter. But the different effects of different substances in transmitting and reflecting heat and light greatly modify this view. One experiment, indeed, of Melloni, already mentioned, would seem to show that light may exist in a condition in which it does not produce heat, which our instruments are able to detect; but some doubt has recently been thrown on the accuracy of this experiment; probably the substances themselves through which the light is transmitted would be found to have been heated.

The recipient body, or that upon which light impinges, seems to exercise nearly as important an influence on our perceptions of light as the emittent body, or that from which the light first proceeds. The recent experiments of Sir John Herschel and Mr. Stokes show that radiant impulses, which, falling on certain bodies, give no effect of light, become luminous when falling on other bodies.

Thus, let ordinary solar light be refracted by a prism (the best material for which is quartz), and the spectrum received on a sheet of paper, or of white porcelain; looking on the paper, the eye detects no light beyond the extreme violet rays. If, therefore, an opaque body be interposed so as just to cut off the whole visible spectrum, the paper would be dark or invisible, with the exception of some slight illumination from light reflected from dust in the air and from surrounding bodies. Substitute for that portion of the paper which was beyond the visible spectrum a piece of glass tinged by the oxide of uranium, and the glass is perfectly visible; so with a bottle of sulphate of quinine, or of the juice of horse-chestnuts, or even paper soaked in these latter solutions. Other substances exhibit in different degrees this effect termed fluorescence; and among the substances which have hitherto been considered perfectly analogous as to their appearance when illuminated, notable differences are discovered. Thus it appears that emanations which give no impression of light to the eye, when impinging on certain bodies, become luminous when impinging on others. We might imagine a room so constructed that such emanations alone are permitted to enter it, which would be dark or light according to the substance with which the walls were coated, though in full daylight the respective coatings of the walls would appear equally white; or, without altering the coating of the walls, the room exposed to one class of rays might be rendered dark by windows which would be transparent to another class.

If, instead of solar light, the electrical light be employed for similar experiments, an equally striking effect can actually be produced. A design, drawn on white paper with a solution of sulphate of quinine and tartaric acid, is invisible by ordinary light, but appears with beautiful distinctness when illuminated by the electric light. Thus, in pronouncing upon a luminous effect, regard must be had to the recipient as well as to the emittent body. That which is, or becomes, light when it falls upon one body is not light when it falls upon another. Probably the retinæ of the eyes of different persons differ to some extent in a similar manner; and the same substance, illuminated by the same spectrum, may present different appearances to different persons, the spectrum appearing more elongated to the one than to the other, so that what is light to the one is darkness to the other. A dependence on the recipient body may also, to a great extent, be predicated of heat. Let two vessels of water, the contents of the one clear and transparent, of the other tinged by some colouring matter, be suspended in a summer's sun; in a very short time a notable difference of temperature will be observed, the coloured having become much hotter than the clear liquid. If the first vessel be placed at a considerable distance from the surface of the earth, and the second near the surface, the difference is still more considerable. Carrying on this experiment, and suspending the first over the top of a high mountain, and the second in a valley, we may obtain so great a difference of temperature, that animals whose organization is suited for the one temperature could not live in the other, and yet both are exposed to the same luminous rays at the same time, and substantially at the same distance from the emittent body—the substance nearer the sun is in fact colder than the more remote. So, with regard to the medium transmitting the influence: a greenhouse may have its temperature considerably varied by changing the glass of which its roof is made.

These effects have an important bearing on certain cosmical questions which have lately been much discussed, and should induce the greatest caution in forming opinions on such subjects as light and heat on the sun's surface, the temperature of the planets, &c. This may depend as much upon their physical constitution as upon their distance from the sun. Indeed, the planet Mars gives us a highly probable argument for this; for, notwithstanding that it is half as far again from the sun as the earth is, the increase of the white tracts at its poles during its winter, and their diminution during its summer, show that the temperature of the surface of this planet oscillates about that of the freezing point of water, as do the analogous zones of our planet. It is true in this we assume that the substance thus changing its state is water, but considering the many close analogies of this planet with the earth, and the identity in appearance of these very effects with what takes place on the earth, it seems not an improbable assumption.

So it by no means necessarily follows, that because Venus is nearer to the sun than the earth, that planet is hotter than our globe. The force emitted by the sun may take a different character at the surface of each different planet, and require different organisms or senses for its appreciation. Myriads of organised beings may exist imperceptible to our vision, even if we were among them; and we might be equally imperceptible to them!

However vain it may be, in the present state of science, to speculate upon such existences, it is equally vain to assume identity or close approxi-

M

mations to our own forms in those beings which may people other worlds. Reasoning from analogy or from final causation, if that be admitted, we may feel convinced that the gorgeous globes of the universe are not unpeopled deserts; but whether the denizens of other worlds are more or less powerful, more or less intelligent, whether they have attributes of a higher or lower class than ourselves, is at present an utterly hopeless guessing.

Specific gravity and intelligence have no necessary connexion. On our own planet five senses, and a mean density equal to that of water, are not invariably associated with intellectual or moral greatness, and the many arguments which have been used to prove that suns and planets other than the earth are uninhabited, or not inhabited by intellectual beings, might, *mutatis mutandis*, equally be used by the denizens of a sun or planet to prove that this world was uninhabited.

Men are too apt, because they are men, because their existence is the one thing of all importance to themselves, to frame schemes of the universe as though it was formed for man alone: painted by an artist of the sun, a man might not represent so prominent an object of creation as he does when represented by his own pencil.

Light was regarded, by what was termed the corpuscular theory, as being in itself matter or a specific fluid emanating from luminous bodies, and producing the effects of sensation by impinging on the retina. This theory gave way to the undulatory one, which is generally adopted in the present day, and which regards light as resulting from the undulation of a specific fluid to which the name of ether has been given, which hypothetic fluid is supposed to pervade the universe, and to permeate the pores of all bodies.

In a Lecture delivered in January 1842, when I first publicly advanced the views advocated in this Essay, I stated that it appeared to me more consistent with known facts to regard light as resulting from a vibration or motion of the molecules of matter itself, rather than from a specific ether pervading it; just as sound is propagated by the vibrations of wood, or as waves are by water. I am not here speaking of the character of the vibrations of light, sound, or water, which are doubtless very different from each other, but am only comparing them so far as they illustrate the propagation of force by motion in the matter itself.

I was not aware, at the time that I first adopted the above view, and brought it forward in my Lectures, that the celebrated Leonard Euler had published a somewhat similar theory; and, though I suggested it without knowing that it had been previously advanced, I should have hesitated in reproducing it had I not found that it was sanctioned by so eminent a mathematician as Euler, who cannot be supposed to have overlooked any irresistible argument against it—the more so in a matter so much controverted and discussed as the undulatory theory of light was in his time.

Although this theory has been considered defective by a philosopher of high repute, I cannot see the force of the arguments by which it has been assailed; and therefore, for the present, though with diffidence, I still adhere to it. The fact itself of the correlation of the different modes of force is to my mind a very cogent argument in favour of their being affections of the same matter; and though electricity, magnetism, and heat might be viewed as produced by undulations of the same either as that by means of which light is supposed to be produced, yet this hypothesis offers greater difficulties with regard to the other affections than with regard to light: many of these difficulties I have already alluded to when treating of electricity; thus conduction and non-conduction are not explained by it; the transmission of electricity through long wires in preference to the air which surrounds them, and which must be at least equally pervaded by the ether, is irreconcilable with such an hypothesis. The phenomena exhibited by these forces afford, as I think, equally strong evidence with those of light, of ordinary matter acting from particle to particle, and having no action at sensible distances. I have already instanced the experiments of Faraday on electrical induction, showing it to be an action of contiguous particles, which are strongly in favour of this view, and many experiments which I have made on the voltaic arc, some of which I have mentioned in this Essay, are, to my mind, confirmatory of it.

If it be admitted that one of the so-called imponderables is a mode of motion, then the fact of its being able to produce the others, and be produced by them, renders it highly difficult to conceive some as molecular motions and others as fluids or undulations of an ether. To the main objection of Dr. Young, that all bodies would have the properties of solar phosphorus if light consisted in the undulations of ordinary matter, it may be answered that so many bodies have this property, and with so great variety in its duration, that *non constat* all may not have it, though for a time so short that the eye cannot detect its duration. M. E. Becquerel has made many experiments which support this view; the fact of the phosphorescence by insolation of a large number of bodies, and the phenomena of fluorescence afford evidence of dense matter being thrown into a state of undulation, or at all events molecularly affected by the impact of light, and is therefore an argument in support of the view which I am advocating. Dr. Young admits that the phenomena of solar phosphorus appear to resemble greatly the sympathetic sounds of musical instruments, which are agitated by other sounds conveyed to them through the air, and I am not aware that he gives any explanation of these effects on the ethereal hypothesis.

Some curious experiments of M. Niepce de St. Victor seem also to present an analogy in luminous phenomena to sympathetic sounds. An engraving which has been kept for some days in the dark is half covered by an opaque screen, and then exposed to the sun; it is then removed from the light, the screen taken away, and the engraving placed opposite, and at a short distance from, photographic paper: an inverted image of that portion of the engraving which has been exposed to the sun is produced on the photographic paper, while the part which had been covered by the screen is not reproduced. If the engraving, after exposure, is allowed to remain in contact with white paper for some hours, and the white paper is then placed upon photographic paper, a faint image of the exposed portion of the engraving is reproduced; similar results are produced by mottled marble exposed to the sun; an invisible tracing on paper by a fluorescent body, sulphate of quinine, is, after insolation, reproduced on photographic paper, &c. Insolated paper retains the power of producing an impression for a very long period, if it is kept in an opaque tube hermetically closed.

It is right to observe that these effects are

supposed by many to be due to chemical emanations proceeding from the substances exposed to the sun, and which are believed to have undergone some chemical change by this exposure; this would still be a molecular effect, but it is desirable to await further experiment before forming a decided opinion.

The analogies in the progression of sound and light are very numerous: each proceed in straight lines, until interrupted; each is reflected in the same manner, the angles of incidence and reflexion being equal; each is alternately nullified and doubled in intensity by interference; each is capable of refraction when passing from media of different density: this last effect of sound, long ago theoretically determined, has been experimentally proved by Mr. Sondhauss, who constructed a lens of films of collodion, which, when filled with carbonic acid, enabled him to hear the ticking of a watch placed at one focus of the lens, the ear of the experimenter being at the opposite focus. The ticking was not heard when the watch was moved aside from the focal point, though it remained at an equal distance from the ear. An experiment of M. Dové seems, indeed, to show an effect of polarisation of sound.

The phenomena presented by heat, viewed according to the dynamic theory, cannot be explained by the motion of an imponderable ether, but involve the molecular actions of ordinary ponderable matter. The doctrine of propagation by undulations of ordinary matter is very generally admitted by those who support the dynamical theory of heat; but the analogies of the phenomena presented by heat and light are so close, that I cannot see how a theory applied to the one agent should not be applicable to the other. When heat is transmitted, reflected, refracted, or polarised, can we view that as an affection of ordinary matter, and when the same effects take place with light, view the phenomena as produced by an imponderable ether, and by that alone?

An objection that immediately occurs to the mind in reference to the ethereal hypothesis of light is, that the most porous bodies are opaque; cork, charcoal, pumice stone, dried and moist wood, &c., all very porous and very light, are all opaque. This objection is not so superficial as it might seem at first sight. The theory which assumes that light is an undulation of an ethereal medium pervading gross matter, assumes the distances between the molecules or atoms of matter to be very great. Matter has been likened by Democritus, and by many modern philosophers, to the starry firmament, in which, though the individual monads are at immense distances from each other, yet they have in the aggregate a character of unity, and are firmly held by attraction in their respective positions and at definite distances.

Now, if matter be built up of separate molecules, then, as far as our knowledge extends, the lightest bodies would be those in which the molecules are at the greatest distances, and those in which any undulation of a pervading medium would be the least interfered with by the separated particles—such bodies should consequently be the most transparent.

If, again, the analogy of the starry firmament held good, in this case an undulation or wave proportioned to the individual monads would be broken up by the number of them, and the very appearance of continuity which results, as in the milky way, from each point of vision being occupied by one of the monads, would show that at some portion of its progress the wave is interrupted by one of them, so that the whole may be viewed in some respect as a sheet of ordinary matter interposed in the ethereal expanse.

Even then, if it be admitted that a highly elastic medium pervades the interspaces, the separate masses as a whole must exercise an important influence on the progress of the wave.

Sound or vibrations of air meeting with a screen, or, as it were, sponge of diffused particles, would be broken up and dispersed by them; but if they be sufficiently continuous to take up the vibration and propagate it themselves, the sound continues comparatively unimpaired.

With regard, however, to liquid and gaseous bodies, there are very great difficulties in viewing them as consisting of separate and distant molecules. If, for instance, we assume with Young that the particles in water are comparatively as distant from each other as 100 men would be if dispersed at equal distances over the surface of England, the distance of these particles, when the water is expanded into steam, would be increased more than forty times, so that the 100 men would be reduced to two, and by further increasing temperature this distance may be indefinitely increased; adding to the effects of temperature rarefaction by the air-pump, we may again increase the distance, so that, if we assume any original distance, we ought, by expansion, to increase it to a point at which the distance between molecule and molecule should become measurable. But no extent of rarefaction, whether by heat or the air-pump, or both, makes the slightest change in the apparent continuity of matter; and gases, I find, retain their peculiar character, as far as a judgment of it can be formed from its effect on the electric spark, throughout any extent of rarefaction which can experimentally be applied to them: thus the electric spark in protoxide of nitrogen, however attenuated, presents a crimson tint, that in carbonic oxide a greenish tint.

Without, however, entering on the metaphysical

enquiry as to the constitution of matter (or whether the atomic philosophers or the followers of Boscovich are right), a problem which probably human appliances will never solve; and even admitting that an ethereal medium, not absolutely imponderable as asserted by many, but of extreme tenuity, pervades matter, still ordinary or non-ethereal matter itself must exercise a most important action upon the transmission of light; and Dr. Young, who opposed the theory of Euler, that light was transmitted by undulations of gross matter itself, just as sound is, was afterwards obliged to call to his assistance the vibrations of the ponderable matter of the refracting media, to explain why rays of all colours were not equally refracted, and other difficulties. One of his arguments in support of the existence of a permeating ether was, 'that a medium resembling in many properties that which has been denominated Ether does exist, is undeniably proved by the phenomena of electricity.' This seems to me, if I may venture to say so of anything proceeding from so eminent a man, scarcely logical: it is supporting one hypothesis by another, and considering that to be proved which its most strenuous advocates admit to be surrounded by very many difficulties.

If it be said that there is not sufficient elasticity in ordinary matter for the transmission of undulations with such velocity as light is known to travel, this may be so if the vibrations be supposed exactly analogous to those of sound; but that molecular motion *can* travel with equal and even greater velocity than light, is shown by the rapidity with which electricity traverses a metal wire where each particle of metal is undoubtedly affected. It has, moreover, been shown by the experiments of Mr. Latimer Clarke upon a length of wire of 760 miles, that whatever be the intensity of electrical currents, they are propagated with the same velocity provided the effects of lateral induction be the same—a striking analogy with one of the effects observed in the propagation of light and sound. The effects observed by MM. Fizeau and Foucault, of the slower progression of light in proportion as the transmitting medium is more dense, seem to me in favour of the view here advocated; as a greater degree of heat would be produced by light in proportion to the density of the medium, force would be thus carried off, and the molecular system disturbed so that the progress of the motion should be more slow; but so many considerations enter into this question, and the phenomena are so extremely complex, that it would be rash to hazard any positive opinion.

Dr. Young ultimately came to the conclusion that it was simplest to consider the ethereal medium, together with the material atoms of the substance, as constituting together a compound medium denser than pure ether, but not more elastic. Ether might thus be viewed as performing the functions which oil does with tracing paper, giving continuity to the particles of gross matter, and in the interplanetary spaces forming itself the medium which transmits the undulations.

Since the period when Huyghens, Euler, and Young, the fathers of the undulatory theory, applied their great minds to this subject, a mass of experimental data has accumulated, all tending to establish the propositions, that whenever matter transmitting or reflecting light undergoes a structural change, the light itself is affected, and that there is a connection or parallelism between the change in the matter and the change in the affection of light, and conversely that light will modify or change the structure of matter and impress its molecules with new characteristics.

Transparency, opacity, refraction, reflection, and colour were phenomena known to the ancients, but sufficient attention does not appear to have been paid by them to the molecular states of the bodies producing these effects; thus the transparency or opacity of a body appears to depend entirely upon its molecular arrangement. If striæ occur in a lens or glass through which objects are viewed, the objects are distorted: increase the number of these striæ, the distortion is so increased that the objects become invisible, and the glass ceases to be transparent, though remaining translucent; but alter completely the molecular structure, as by slow solidification, and it becomes opaque. Take, again, an example of a liquid and a gas: a solution of soap is transparent, air is transparent, but agitate them together so as to form a froth or lather, and this, though consisting of two transparent bodies, is opaque; and the reflection of light from the surface of these bodies, when so intermixed, is strikingly different from its reflection before mixture, in the one case giving to the eye a mere general effect of whiteness, in the other the images of objects in their proper shapes and colours.

To take a more refined instance: nitrogen is perfectly colourless, oxygen is perfectly colourless, but chemically united in certain proportions they form nitrous acid, a gas which has a deep orange-brown colour. I know not how the colour of this gas, or of such gases as chlorine or vapour of iodine, can be accounted for by the ethereal hypothesis, without calling in aid molecular affections of the matter of these gases.

Colour in many instances depends upon the thickness of the plate or film of transparent matter upon which light is incident; as in all those cases which are termed the colours of thin plates, of which the soap bubble affords a beautiful instance.

When we arrive at the more recent discoveries of double refraction and polarisation, the effects of light are found to trace out as it were the structure of the matter affected, and the crystalline form of a body can be determined by the effects which a minute portion of it exercises on a ray of light.

Let a piece of good glass be placed in what is called a polariscope (i.e. an instrument in which light that has undergone polarisation is transmitted through the substance to be examined, and the emergent light is afterwards submitted to another substance capable of polarising light, or, as it is termed, an analyser), no change in effect will be observed. Remove the glass, heat it and suddenly or quickly cool it so as to render it unannealed, in which state its molecules are in a state of tension or strain, and the glass highly brittle, and on replacing it in the polariscope, a beautiful series of colours is perceptible. Instead of subjecting the glass to heat and sudden cooling, let it be bent or strained by mechanical pressure, and the colours will be equally visible, modified according to the direction of the flexure, and indicating by their course the curves where the molecular state has been changed by pressure. So if tough glue be elongated and allowed to cool in a stretched state, it doubly refracts light, and the colours are shown as in the instance of glass.

Submit a series of crystals to the same examination, and different figures will be formed by different crystals, bearing a constant and definite relation to the structure of the particular crystal examined, and to the direction in which, with reference to crystalline form, the ray crosses the crystal.

In the crystallised salts of paratartaric acid, M. Pasteur noticed two sets of crystals which were hemihedral in opposite directions, i. e. the crystals of one set were to those of the other as to their own image reflected in a mirror; on making a separate solution of each of these classes of crystals, he found that the solution of the one class rotated the plane of polarisation to the right, while that of the other class rotated it to the left, and that a mixture in proper proportions of the two solutions produced no deviation in the plane of polarisation. Yet all these three solutions are what is termed isomeric, that is, have as far as can be discovered the same chemical constitution.

In the above, and in innumerable other cases, it is seen that an alteration in the structure of a transparent substance alters the character and effects of the transmitted light. The phenomena of photography prove that light alters the structure of matter submitted to it; with regard even to vision itself, the persistence of images on the retina of the eye would seem to show that its strucure is changed by the impact of light, the luminous impressions being as it were branded on the retina, and the memory of the vision being the scar of such brand. The science of photography has reference mainly to solid substances, yet there are many instances of liquid and gaseous bodies being changed by the action of light: thus hydrocyanic acid, a liquid, undergoes a chemical change and deposits a solid carbonaceous compound by the action of light. Chlorine and hydrogen gases, when mixed and preserved in darkness, do not unite, but when exposed to light rapidly combine, forming hydrochloric acid.

The above facts—and many others might have been given—go far to connect light with motion of ordinary matter, and to show that many of the evidences which our senses receive of the existence of light result from changes in matter itself. When the matter is in the solid state, these changes are more or less permanent; when in the liquid or gaseous state, they are temporary in the greater number of instances, unless there be some chemical change effected, which is, as it were, seized upon during its occurrence, and a resulting compound formed, which is more stable than the original compound or mixture.

I might weary my reader with examples, showing that, in every case which we can trace out, the effects of light are changed by any and every

N

change of structure, and that light has a definite connection with the structure of the bodies affected by it. I cannot but think that it is a strong assumption to regard ether, a purely hypothetical creation, as changing its elasticity for each change of structure, and to regard it as penetrating the pores of bodies of whose porosity we have in many cases no proof; the which pores must, moreover, have a definite and peculiar communication, also assumed for the purpose of the theory.

Ether is a most convenient medium for hypothesis: thus, if to account for a given phenomenon the hypothesis requires that the ether be more elastic, it is said to be more elastic; if more dense, it is said to be more dense; if it be required by hypothesis to be less elastic, it is pronounced to be less elastic; and so on.

The advocates of the ethereal hypothesis certainly have this advantage, that the ether, being hypothetical, can have its characters modified or changed without any possibility of disproof either of its existence or modifications.

It may be that the refined mathematical labours on light, as on electricity, have given an undue and adventitious strength to the hypotheses on which they are based.

An objection to which the view I have been advocating is open, and at first sight a formidable one, is the necessity involved in it of an universal plenum; for if light, heat, electricity, &c., be affections of ordinary matter, then matter must be supposed to be everywhere where these phenomena are apparent, and consequently there can be no vacuum.

These forces are transmitted through what are called vacua, or through interplanetary spaces, where matter, if it exist, must be in a highly attenuated state.

It may be safely stated that hitherto all attempts at procuring a perfect vacuum have failed. The ordinary air-pump gives us only highly rarefied air; and, by the principle of construction, even of the best, the operation depends upon the indefinite expansion of the volume of air in the receiver; even in the vacuum which is formed in this, so great is the tendency of matter to fill up space, that I have observed distilled water contained in a vessel within the exhausted receiver of a good air-pump has a taste of tallow, derived from the grease, or an essential oil contained in it, which is used to form an air-tight junction between the edges of the receiver and the pump-plate.

The Torricellian vacuum, or that of the ordinary barometer, is filled with the vapour of mercury; but it might be worth the trouble to ascertain what would be the effect of a good Torricellian vacuum, when the mercury in the tube is frozen, which might, without much difficulty, be now effected by the use of solid carbonic acid and ether; the only probable difficulty would be the different rates of contraction of mercury and glass, at such a degree of cold, and more particularly the contraction of mercury at the period of its solidification. Davy, indeed, endeavoured to form a vacuum, in a somewhat similar manner, over fused tin, with but partial success; he also made many other attempts to obtain a perfect vacuum; his main object being to ascertain what would be the effect of electricity across empty space: he admits that he could not succeed in procuring a vacuum, but found electricity much less readily conducted or transmitted by the best vacuum he could procure than by the ordinary Boylean vacuum.

Morgan found no conduction by a good Torricellian vacuum; and, although Davy does not seem to place much reliance on Morgan's experiments, there was one point in which they were less liable to error than those of Davy. Morgan, whose experiments seem to have been carefully conducted, operated with hermetically-sealed glass tubes and by induced electricity, while Davy sealed a platinum wire into the extremity of the tube in which he sought to produce a vacuum. I have found in very numerous experiments which I made to exclude air from water, that platinum wires, most carefully sealed into glass, allow liquids to pass between them and the glass; and this gives some reason to believe that gases may equally pass through; indeed, I have observed such effect in the gas battery when it has been in action for a long period. Davy supposed that the particles of bodies may be detached, and so produce electrical effects in a vacuum; and such effects would more readily take place in his experiments, where a wire projected into the exhausted space, than in Morgan's, where the induced electricity was diffused over the surface of the glass.

M. Masson found that the barometric vacuum does not conduct a current of electricity, or even a discharge, unless the tension is considerable and sufficient to detach particles from the electrodes; and by adopting a plan of Dr. Andrews, viz. absorbing carbonic acid by potash, Mr. Gassiot has succeeded in forming vacua across which the powerful discharge from the Rhumkorf coil will not pass.

The odour which many metals, such as iron, tin, and zinc emit, and the so-called thermographic radiations, can hardly be explained upon any other theory than the evaporation of an infinitesimally small portion of the metal itself.

So universal is the tendency of matter to diffuse itself into space, that it gave rise to the old saying that nature abhors a vacuum; an aphorism which, though cavilled at and ridiculed by the self-sufficiency of some modern philosophers, contains in a terse though somewhat metaphorical form of expression, a comprehensive truth, and evinces a large extent of observation in those who, with few of the advantages which we possess, first generalised by this sentence the facts of which they had become cognisant.

It has been argued that if matter were capable of infinite divisibility, the earth's atmosphere would have no limit, and that consequently portions of it would exist at points of space where the attraction of the sun and planets would be greater than that of the earth, and whence it would fly off to those bodies and form atmospheres around them. This was supposed to be negatived by the argument of the well-known paper of Dr. Wollaston; in which, from the absence of notable refraction near the margin of the sun and of the planet Jupiter, he considered himself entitled to conclude that the expansion of the earth's atmosphere had a definite limit, and was balanced at a certain point by gravitation: this deduction has been shown to be inconclusive by Dr. Whewell, and has also been impugned upon other grounds by Dr. Wilson. There is a point not adverted to in these papers, and which Wollaston does not seem to have considered, viz. that there is no evidence that the apparent discs of the sun and of Jupiter show us their real discs or bodies. Sir W. Herschel regards the margin of the visible discs as that of clouds or a peculiar state of atmosphere; and the rapidly changing character of the apparent surfaces renders some such conclusion necessary. If this be so, refraction of an occulted star could not be detected—at all events, in the denser portion of the atmosphere.

Sir W. Herschel's observations go to prove that the sun and Jupiter have dense atmospheres, while Wollaston's were believed to prove that they have no appreciable atmospheres.

If it be admitted, or considered proved, that the sun and planets have atmospheres—and little doubt now exists on this point—then the grounds upon which Wollaston founded his arguments are untenable; and there appears no reason why the atmosphere of the different planets should not be, with reference to each other, in a state of equilibrium. Ether, or the highly-attenuated matter existing in the interplanetary spaces, being an expansion of some or all of these atmospheres, or of the more volatile portions of them, would thus furnish matter for the transmission of the modes of motion which we call light, heat, &c.; and possibly minute portions of these atmospheres may, by gradual changes, pass from planet to planet, forming a link of material communication between the distant monads of the universe.

The assumption of the universal presence of matter is common to the theory of the transmission of light by the undulations of ordinary matter and to the other two theories, which equally suppose the non-existence of a vacuum; for, according to the emissive or corpuscular theory, the vacuum is filled by the matter itself, of light, heat, &c.; according to the ethereal it is filled by the all-penetrating ether. Of the existence of matter in the interplanetary spaces we have some evidence in the diminishing periods of comets; and where, from its highly attenuated state, the character of the medium by which the forces are conveyed cannot be tested, the term ether is a most appropriate generic name for such medium.

Newton has some curious passages on the subject matter of light. In the 'Queries to the Optics' he says:—

'Are not gross bodies and light convertible into one another, and may not bodies receive much of their activity from the particles of light which enter their composition? * * * The changing of bodies into light and light into bodies is very conformable to the course of nature, which seems delighted with transmutations. Water, which is a very fluid, tasteless salt, she changes by heat into vapour, which is a sort of air, and by cold into ice, which is a hard, pellucid, brittle, fusible stone, and this stone returns into water by heat, and vapour returns into water by cold. * * * And, among such various and strange transmutations, why may not nature change bodies into light, and light into bodies?'

Newton has here seemingly in his mind the emissive theory of light; but the passages might be applied to either theory; the analogy he saw in the change of state of matter, as in ice, water, and vapour, with the hypothetic change into light, is very striking, and would seem to show that he regarded the change or transmutation of which he speaks as one analogous to the known changes of state, or consistence, in ordinary matter.

The difference between the view which I am advocating and that of the ethereal theory as generally enunciated is, that the matter which in the interplanetary spaces serves as the means of transmitting by its undulations light and heat, I should regard as possessing the qualities of ordinary, or as it has sometimes been called gross, matter, and particularly weight; though, from its extreme rarefaction, it would manifest these properties in an indefinitely small degree; whilst, near to or on the surface of the earth, that matter attains a density cognisable by our means of experiment, and the dense matter is itself, in great part, the conveyer of the undulations in which these agents consist. Doubtless, in very many of the forms which matter assumes it is porous, and pervaded by more volatile essences, which may differ as much in kind as matter does. In these cases a composite medium, such as that indicated by Dr. Young, would result; but even on such a supposition, the denser matter would probably exercise the more important influence on the undulations. Returning to the somewhat strained hypothesis, that the particles of dense matter in a so-called solid are as distant as the stars in heaven, still a certain depth or thickness of such solid would present at every point of space a particle or rock in the successive progress of a wave, which particles, to carry on the movement, must vibrate in unison with it.

At the utmost, our assumption, on the one hand, is, that wherever light, heat, &c., exist, ordinary matter exists, though it may be so attenuated that we cannot recognise it by the tests of other forces, such as gravitation; and that to the expansibility of matter no limit can be assigned. On the other hand, a specific matter without weight must be assumed, of the existence of which there is no evidence, but in the phenomena for the explanation of which its existence is supposed. To account for the phenomena the ether is assumed, and to prove the existence of the ether the phenomena are cited. For these reasons, and others above given, I think that the assumption of the universality of ordinary matter is the least gratuitous.

Οὐδέν τι τοῦ παντὸς κενὸν πέλει οὐδὲ περισσόν.

A question has often occurred to me and possibly to others: Is the continuance of a luminous impulse in the interplanetary spaces perpetual, or does it after a certain distance dissipate itself and become lost as light—I do not mean by mere divergence directly as the squares of the distances it travels, but does the physical impulse itself lose force as it proceeds? Upon the view I have advocated, and indeed upon any undulatory hypothesis, there must be some resistance to its progress; and unless the matter or ether in the interplanetary spaces be infinitely elastic, and there be no lateral action of a ray of light, there must be some loss. That it is exceedingly minute is proved by the distance light travels. Stars whose parallax is ascertained are at such a distance from the earth that their light, travelling at the rate of 192,500 miles in a second, takes more than ten years to reach the earth; so that we see them as they existed ten years ago. The distance of most visible stars is probably far greater than this, and yet their brilliance is great, and increases when their rays are collected by the telescope in proportion *ceteris paribus* to the area of the object-glass or speculum. There is, however, an argument of a somewhat speculative character, by which light would seem to be lost or transformed into some other force in the interplanetary spaces.

Every increase of space-penetrating power in the telescope gives us a new field of visible stars. If this expansion of the stellar universe go on indefinitely and no light be lost, then, assuming the fixed stars to be of an average equal brightness with our sun, and no light lost other than by divergence, the night ought to be equally luminous with the day; for though the light from each point diminishes in intensity as the square of the distance, the number of luminous points would increase as the square of the distance, and thus fill up the whole space around us; and if every point of space is occupied by an equally brilliant point of light, the distance of the points from us becomes immaterial. The loss of light intercepted by stellar bodies would make no difference in the total quantity of light, for each of these would yield from its own self-luminosity at least as much light as it intercepted. Light may, however, be intercepted by non-luminous bodies, such as planets; but, making every allowance for these, it is difficult to understand why we get so little light at night from the stellar universe, without assuming that some light is lost in its progress through space —not lost absolutely, for that would be an annihilation of force—but converted into some other mode of motion.

It may be objected that this hypothesis assumes the stellar universe to be illimitable: if pushed to its extreme so as to make the light of night equal that of day provided no stellar light be lost, it does make this assumption; but even this is a far more rational assumption to make than that the stellar universe is limited. Our experience gives no indication of a limit; each improvement in telescopic power gives us new realms of stars or of nebulæ, which, if not stellar clusters, are at all events self-luminous matter; and if we assume a limit, what is it? We cannot conceive a physical boundary, for then immediately comes the question, what bounds the boundary? and to suppose the stellar universe to be bounded by infinite space or by infinite chaos, that is to say, to suppose a spot—for it would then become so—of matter in definite forms, with definite forces, and probably teeming with definite organic beings, plunged in a universe of nothing, is to my mind at least far more unphilosophical than to suppose a boundless universe of matter existing in forms and actions more or less analogous to those which, as far as our examination goes, pervade space. But without speculating on topics in which the mind loses itself, it may not unreasonably be expected that a greater amount of light would reach us from the surrounding self-luminous spheres were not some portion lost as light, by its action on the medium which conveys the impulses. What force this becomes, or what it effects, it would be vain to speculate upon.

MAGNETISM.

MAGNETISM, as was proved by the important discovery of Faraday, will produce *electricity*, but with this peculiarity—that in itself it is static; and, therefore, to produce a dynamic force, motion must be superadded to it: it is, in fact, directive, not motive, altering the direction of other forces, but not, in strictness, initiating them. It is difficult to convey a definite notion of the force of magnetism, and of the mode in which it affects other forces. The following illustration may give a rude idea of magnetic polarity. Suppose a number of wind-vanes, say of the shape of arrows, with the spindles on which they revolve arranged in a row, but the vanes pointing in various directions: a wind blowing from the same point with an uniform velocity will instantly arrange these vanes in a definite direction, the arrow-heads or narrow parts pointing one way, the swallow-tails or broad parts another. If they be delicately suspended on their spindles, a very gentle breeze will so arrange them, and a very gentle breeze will again deflect them; or, if the wind cease, and they have been originally subject to other forces, such as gravity from unequal suspension, they will return to irregular positions, themselves creating a slight breeze by their return. Such a state of things will represent the state of the molecules of soft iron; electricity acting on them—not indeed in straight lines, but in a definite direction—produces a polar arrangement, which they will lose as soon as the dynamic inducing force is removed.

Let us now suppose the vanes, instead of turning easily, to be more stiffly fixed to the axles, so as to be turned with difficulty: it will require a stronger wind to move them and arrange them definitely; but when so arranged, they will retain their position; and should a gentle breeze spring up in another direction, it will not alter their position, but will. itself be definitely deflected. Should the conditions of force and stability be intermediate, both the breeze and the vanes will be slightly deflected; or, if there be no breeze, and the spindles be all moved in any direction, preserving their linear relation, they will themselves create a breeze. Thus it is with the molecules of hard iron or steel in permanent magnets; they are polarised with greater difficulty than those of soft iron, but, when so polarised, they cannot be affected by a feeble current of electricity. Again, if the magnets be moved, they themselves originate a current of electricity; and, lastly, the magnetic polarity and the electric current may be both mutually affected, if the degrees of mobility and stability be intermediate.

The above instance will, of course, be taken only as an approximation, and not as binding me to any closer analogy than is generally expected of a mechanical illustration. It is difficult to convey by words a definite idea of the dual or antithetic character of force involved in the term polarity. The illustration I have employed may, I hope, somewhat aid in elucidating the manner in which magnetism acts on the other dynamic forces; i. e., definitely directing them, but not initiating them, except while in motion.

Magnets being moved in the direction of lines, joining their poles, produce electrical currents in such neighbouring bodies as are conductors of electricity, in directions transverse to the line of motion; and if the direction of motion or the position of the magnetic poles be reversed, the current of electricity flows in a reverse direction. So if the magnet be stationary, conducting bodies moved across any of the lines of magnetic force (i. e. lines in the direction of which the mutual action of the poles of the magnet would place minute portions of iron) have currents of electricity developed in them, the direction of which is dependent upon that of the motion of the substance with reference to the magnetic poles. Thus, as bodies affected by an electrical current are definitely moved by a magnet in proximity to them, so conversely bodies moved near a magnet have an electrical current developed in them. Magnetism can, then, through the medium of electricity, produce *heat, light*, and *chemical affinity*. *Motion* it can directly produce under the above conditions; i.e. a magnet being itself moved will move other ferreous bodies: these will acquire a static condition of equilibrium, and be again moved when the magnet is also moved. By motion or arrested motion only, could the phenomena of magnetism ever have become known to us. A magnet, however powerful, might rest for ever unnoticed and unknown, unless it were moved near to iron, or iron moved near to it, so as to come within the sphere of its attraction.

But even with other than either magnetic or electrified substances, all bodies will be moved when placed near the poles of very powerful magnets — some taking a position axially, or in the line from pole to pole of the magnet; others equatorially, or in a direction transverse to that line—the former being attracted, the latter apparently repelled, by the poles of the magnet. These effects, according to the views of Faraday, show a generic difference between the two classes of bodies, magnetics and diamagnetics; according to others, a difference of degree or a resultant of magnetic action; the less magnetic substance

o

being forced into a transverse position by the magnetisation of the more magnetic medium which surrounds it.

According to the view given above, magnetism may be produced by the other forces, just as the vanes in the instance given are definitely deflected, but cannot produce them except when in motion : motion, therefore, is to be regarded in this case as the initiative force. Magnetism will, however, directly affect the other forces—light, heat, and chemical affinity, and change their direction or mode of action, or, at all events, will so affect matter subjected to these forces, that their direction is changed. Since these lectures were delivered, Faraday has discovered a remarkable effect of the magnetic force in occasioning the deflection of a ray of polarised light.

If a ray of polarised light pass through water, or through any transparent liquid or solid which does not alter or turn aside the plane of polarisation, and the column, say of water, through which it passes be subjected to the action of a powerful magnet, the line of magnetic force, or that which would unite the poles of the magnet, being in the same direction as the ray of polarised light, the water acquires, with reference to the light, similar, though not quite identical, properties to oil of turpentine — the plane of polarisation is rotated, and the direction of this rotation is changed by changing the direction of the magnetic force: thus, if we suppose a polarised ray to pass first in its course the north pole of the magnet, then between that and the south pole it will be deflected, or curved to the right; while if it meet the south pole first in its course, it will, in its journey between that and the north pole, be turned to the left. If the substance through which the ray is transmitted be of itself capable of deflecting the plane of polarisation, as, for instance, oil of turpentine, then the magnetic influence will increase or diminish this rotation, according to its direction. A similar effect to this is observed with polarised heat when the medium through which it is transmitted is subjected to magnetic influence.

Whether this effect of magnetism is rightly termed an effect upon light and heat, or is a molecular change of the matter transmitting the light and heat, is a question the resolution of which must be left to the future; at present, the answer to it would depend upon the theory we adopt. If the view of light and heat which I have stated be adopted, then we may fairly say that magnetism, in these experiments, directly affects the other forces; for light and heat being, according to that view, motions of ordinary matter, magnetism, in affecting these movements, affects the forces which occasion them. If, however, the other theories be adhered to, it would be more consistent with the facts to view these results as exhibiting an action upon the matter itself, and the heat and light as secondarily affected.

When substances are *undergoing* chemical changes, and a magnet is brought near them, the direction or lines of action of the chemical force will be changed. There are many old experiments which probably depended on this effect, but which were erroneously considered to prove that permanent magnetism could produce or increase chemical action: these have been extended and explained by Mr. Hunt and Mr. Wartmann, and are now better understood.

The above cases are applicable to the subject of the present Essay, inasmuch as they show a relation to exist between magnetic and the other forces, which relation is, in all probability, reciprocal; but in these cases there is not a production of light, heat, or chemical affinity, by magnetism, but a change in their direction or mode of action.

There is, however, that which may be viewed as a dynamic condition of magnetism; i.e. its condition at the commencement and the termination, or during the increment or decrement of its development. While iron or steel is being rendered magnetic, and as it progresses from its non-magnetic to its maximum magnetic state, or recedes from its maximum to zero, it exhibits a dynamic force; the molecules are, it may be inferred, in motion. Similar effects can then be produced to those which are produced by a magnet whilst in motion.

An experiment which I published in 1845 tends, I think, to illustrate this, and in some degree to show the character of the motion impressed upon the molecules of a magnetic metal at the period of magnetisation. A tube filled with the liquid in which magnetic oxide of iron had been prepared, and terminated at each end by plates of glass, is surrounded by a coil of coated wire. To a spectator looking through this tube a flash of light is perceptible whenever the coil is electrised, and less light is transmitted when the electrical current ceases, showing a symmetrical arrangement of the minute particles of magnetic oxide while under the magnetic influence.

In this experiment it should be borne in mind, that the particles of oxide of iron are not shaped by the hand of man, as would be the case with iron filings, or similar minute portions of magnetic matter, but being chemically precipitated, are of the form given to them by nature.

While magnetism is in the state of change above described, it will produce the other forces; but it may be said, while magnetism is thus progressive, some other force is acting on it, and therefore it does not initiate: this is true, but the same may be said of all the other forces; they have no commencement that we can trace. We must ever refer them back to some antecedent force equal in amount to that produced, and therefore the word initiation cannot in strictness apply, but must only be taken as signifying the force selected as the first: this is another reason why the idea of abstract causation is inapplicable to physical production. To this point I shall again advert.

Electricity may thus be produced directly by magnetism, either when the magnet as a mass is in motion, or when its magnetism is commencing, increasing, decreasing, or ceasing; and heat may similarly be directly produced by magnetism. I have, since the first edition of this Essay was published, communicated to the Royal Society a paper by which I think I have satisfactorily proved, that whenever any metal susceptible of magnetism is magnetised or demagnetised, its temperature is raised. This was shown, first, by subjecting a bar of iron, nickel, or cobalt to the influence of a powerful electro-magnet, which was rapidly magnetised and demagnetised in reverse directions, the electro-magnet itself being kept cool by cisterns of water, so that the magnetic metal subjected to the influence of magnetism was raised to a higher temperature than the electromagnet itself, and could not, therefore, have acquired its increased temperature by conduction or radiation of heat from the electro-magnet; and secondly, by rotating a permanent steel magnet with its poles opposite to a bar of iron, a thermo-electric pile being placed opposite the latter.

Dr. Maggi covered a plate of homogeneous soft iron with a thin coating of wax mixed with oil, a tube traversed the centre through which the vapour of boiling water was passed. The plate was made to rest on the poles of an electro-magnet, with card interposed. When the iron is not magnetised, the melted wax assumes a circular form, the tube occupying the centre, but when the electro-magnet is put in action, the curve marking the boundary of the melted substance changes its form and becomes elongated in a direction transverse to the line joining the poles, showing that the conducting power of the iron for heat is changed by magnetisation.

Thus we get heat produced by magnetism and the conduction of heat altered by it in a direction having a definite relation to the direction of the magnetism. Is it necessary to call in aid ether or the substance 'caloric' to explain these results? is it not more rational to regard the calorific effects as changes in the molecular arrangements of the matter subjected to magnetism?

There is some probability that magnetism, in the dynamic state, either when the magnet is in motion, or when the magnetic intensity is varying, will also directly produce chemical affinity and light, though up to the present time, such has not been proved to be the case; the reciprocal effect, also, of the direct production of magnetism by light and heat has not yet been experimentally established.

I have used, in contradistinction, the terms dynamic and static to represent the different states of magnetism. The applications I have made of these terms may be open to some exception, but I know of no other words which will so nearly express my meaning.

The static condition of magnetism resembles the static condition of other forces: such as the state of tension existing in the beam and cord of a balance, or in a charged Leyden phial. The old definition of force was, that which caused change in motion; and yet even this definition presents a difficulty: in a case of static equilibrium, such, for instance, as that which obtains in the two arms of a balance, we get the idea of force without any palpable apparent motion: whether there be really an absence of motion may be a doubtful question, as such absence would involve in this case perfect elasticity, and, in all other cases, a stability which, in a long course of time, nature generally negatives, showing, as I believe, an inseparable connection of motion with matter, and an impossibility of a perfectly immobile or durable state. So with magnetism; I believe no magnet can exist in an absolutely stable state, though the duration of its stability will be proportionate to its original resistance to assuming a polarised condition. This however, must be taken merely as a matter of opinion: we have, in support of it, the general fact that magnets do deteriorate in the course of years; and we have the further general fact of the instability, or fluxional state, of all nature, when we have an opportunity of fairly investigating it at different and remote periods: in many cases, however, the action is so slow that the changes escape human observation, and, until this can be brought to bear over a proportionate period of time, the proposition cannot be said to be experimentally or inductively proved, but must be left to the mental conviction of those who examine it by the light of already acknowledged facts.

All cases of static force present the same difficulty: thus, two springs pressing against each other would be said to be exercising force; and yet there is no resulting action, no heat, no light, &c.

So if gas be compressed by a piston, at the time of compression heat is given off; but when this is abstracted, although the pressure continues, no further heat is eliminated. Thus, by an equilibrium produced by opposing forces, motion is locked up, or in abeyance, as it were, and may be again developed when the forces are relieved from the tension. But in the first instance, in producing the state of tension, force has to be employed; and as we have said in treating of mechanical force, so with the other forces the original change which disturbs equilibrium produces other changes which go on without end. Thus, by the act of charging a Leyden phial, the cylinder, the rubber, and the adjoining portions of the electrical machine have each and all their states changed, and thence produce changes in surrounding bodies *ad infinitum*; when the jar is discharged, converse changes are again produced.

As with heat, light, and electricity, the daily accumulating observations tend to show that each change in the phenomena to which these names are given is accompanied by a change either temporary or permanent in the matter affected by them; so many experiments on magnetism have connected magnetic phenomena with a molecular change in the subject matter. Thus M. Wertheim has shown that the elasticity of iron and steel is altered by magnetisation; the co-efficient of elasticity in iron being temporarily, in steel permanently diminished.

He has also examined the effects of torsion upon magnetised iron, and concludes, from his experiments, that in a bar of iron arrived at a state of magnetic equilibrium, temporary torsion diminishes the magnetism, and that the untwisting or return to its primitive state restores the original degree of magnetisation.

M. Guillemin observed that a bar slightly curved by its own weight is straightened by being magnetised. Mr. Page and Mr. Marrion discovered that a sound is emitted when iron or steel is rapidly magnetised or demagnetised; and Mr. Joule found that a bar of iron is slightly elongated by magnetisation.

Again, with regard to diamagnetic bodies. M. Matteucci found that the mechanical compression of glass altered the rotatory power of magnetism upon a ray of polarised light which the glass transmitted. He further considered that a change took place in the temper of portions of glass which he submitted to the influence of powerful magnets.

The same arguments which have been submitted to the reader as to the other affections of matter being modes of molecular motion, are therefore equally applicable to magnetism.

CHEMICAL AFFINITY.

CHEMICAL AFFINITY, or the force by which dissimilar bodies tend to unite and form compounds differing generally in character from their constituents, is that mode of force of which the human mind has hitherto formed the least definite idea. The word itself—*affinity*—is ill chosen, its meaning, in this instance, bearing no analogy to its ordinary sense; and the mode of its action is conveyed by certain conventional expressions, no dynamic theory of it worthy of attention having been adopted. Its action so modifies and alters the character of matter, that the changes it induces have acquired, not perhaps very logically, a generic contradistinction from other material changes, and we thus use, as contradistinguished, the terms physical and chemical.

The main distinction between chemical affinity and physical attraction or aggregation, is the difference of character of the chemical compound from its components. This is, however, but a vague line of demarcation; in many cases, which would be classed by all as chemical actions, the change of character is but slight; in others, as in the effects of neutralisation, the difference of character would be a result which would equally follow from physical attraction of dissimilar substances, the previous characters of the constituents depending upon this very attraction or affinity: thus an acid corrodes because it tends to unite with another body; when united, its corrosive power, i. e. its tendency to unite, being satiated, it cannot, so to speak, be further attracted, and it necessarily loses its corrosive power. But there are other cases where no such result could *à priori* be anticipated, as where the attraction or combining tendency of the compound is higher than that of any of its constituents; thus, who could, by physical reasoning, anticipate a substance like nitric acid from the combination of nitrogen and oxygen?

The nearest approach, perhaps, that we can form to a comprehension of chemical action, is by regarding it (vaguely perhaps) as a molecular attraction or motion. It will directly produce motion of definite masses, by the resultant of the molecular changes it induces: thus, the projectile effects of gunpowder may be cited as familiar instances of motion produced by chemical action. It may be a question whether, in this case, the force which occasions the motion of the mass is a conversion of the force of chemical affinity, or whether it is not, rather, a liberation of other forces existing in a state of static equilibrium, and having been brought into such state by previous chemical actions; but, at all events, through the medium of electricity chemical affinity may be directly and quantitatively converted into the other modes of force. By chemical affinity, then, we can directly produce electricity; this latter force was, indeed, said by Davy to be chemical affinity acting on masses: it appears, rather to be chemical affinity acting in a definite direction through a chain of particles; but by no definition can the exact relation of chemical affinity and electricity be expressed; for the latter, however closely related to the former, yet exists where the former does not, as in a metallic wire, which, when electrified, or conducting electricity, is, nevertheless, not chemically altered, or, at least, not known to be chemically altered.

Volta, the antitype of Prometheus, first enabled us definitely to relate the forces of chemistry and electricity. When two dissimilar metals in contact are immersed in a liquid belonging to a certain class, and capable of acting chemically on one of them, what is termed a voltaic circuit is formed, and, by the chemical action, that peculiar mode of force called an electric current is generated, which circulates from metal to metal, across the liquid, and through the points of contact.

Let us take, as an instance of the conversion of chemical force into electrical, the following, which

I made known some years ago. If gold be immersed in hydrochloric acid, no chemical action takes place. If gold be immersed in nitric acid, no chemical action takes place; but mix the two acids, and the immersed gold is chemically attacked and dissolved: this is an ordinary chemical action, the result of a double chemical affinity. In hydrochloric acid, which is composed of chlorine and hydrogen, the affinity of chlorine for gold being less than its affinity for hydrogen, no change takes place; but when the nitric acid is added, this latter containing a great quantity of oxygen in a state of feeble combination, the affinity of oxygen for hydrogen opposes that of hydrogen for chlorine, and then the affinity of the latter for gold is enabled to act, the gold combines with the chlorine and chloride of gold remains in solution in the liquid. Now, in order to exhibit this chemical force in the form of electrical force, instead of mixing the liquids, place them in separate vessels or compartments, but so that they may be in contact, which may be effected by having a porous material, such as unglazed porcelain, amianthus, &c., between them. Immerse in each of these liquids a strip or wire of gold: as long as these pieces of gold remain separated, no chemical or electrical effect takes place; but the instant they are brought into metallic contact, either immediately or by connecting each with the same metallic wire, chemical action takes place — the gold in the hydrochloric acid is dissolved, electrical action also takes place, the nitric acid is deoxidised by the transferred hydrogen, and a current of electricity may be detected in the metals, or connecting metal, by the application of a galvanometer or any instrument appropriate for detecting such effect.

There are few, if any, chemical actions which cannot be experimentally made to produce electricity: the oxidation of metals, the burning of combustibles, the combination of oxygen and hydrogen, &c., may all be made sources of electricity. The common mode in which the electricity of the voltaic battery is generated is by the chemical action of water upon zinc; this action is increased by adding certain acids to the water, which enable it to act more powerfully upon the zinc, or in some cases act themselves upon it; and one of the most powerful chemical actions known —that of nitric acid upon oxidable metals—is that which produces the most powerful voltaic battery, a combination which I made known in the year 1839: indeed, we may safely say, that when the chemical force is utilised, or not wasted, but all converted into electrical force, the more powerful the chemical action, the more powerful is the electrical action which results.

If, instead of employing manufactured products

or educts, such as zinc and acids, we could realise as electricity the whole of the chemical force which is active in the combustion of cheap and abundant raw materials, such as coal, wood, fat, &c., with air or water, we should obtain one of the greatest practical desiderata, and have at our command a mechanical power in every respect superior in its applicability to the steam engine.

I have shown that the flame of the common blowpipe gives rise to a very marked electrical current, capable not only of affecting the galvanometer, but of producing chemical decomposition: two plates or coils of platinum are placed, the one in the portion of the flame near the orifice of the jet, or at the points where combustion commences, the other in the full yellow flame where combustion is at its maximum; this latter should be kept cool, to enable a thermo-electric current, which is produced by the different temperature of the platinum plates, to co-operate with the flame current; wires attached to the plates of platinum form the terminals or poles. By a row of jets a flame battery may be formed, yielding increased effects; but in these experiments, though theoretically interesting, so small a fraction of the power, actually at work in the combustion, has been thrown into an electrical form, that there is no immediate promise of a practical result.

The quantity of the electrical current, as measured by the quantity of matter it acts upon in its different phenomenal effects, is proportionate to the quantity of chemical action which generated it; and its intensity, or power of overcoming resistance, is also proportionate to the intensity of chemical affinity when a single voltaic pair is employed, or to the number of reduplications when the well-known instrument called the voltaic battery is used.

The mode in which the voltaic current is increased in intensity by these reduplications, is in itself a striking instance of the mutual relations and dynamic analogies of different forces. Let a plate of zinc or other metal possessing a strong affinity for oxygen, and another of platinum or other metal possessing little or no affinity for oxygen, be partially immersed in a vessel, A, containing dilute nitric acid, but not in contact with each other; let platinum wires touching each of these plates have their extremities immersed in another vessel, B, containing also dilute nitric acid: as the acid in vessel A is decomposed, by the chemical affinity of the zinc for the oxygen of the acid, the acid in vessel B is also decomposed, oxygen appearing at the extremity of the wire which is connected with the platinum: the chemical power is conveyed or transferred through the wires, and, abstracting certain local effects, for every unit of oxygen which combines with the zinc in the one vessel, a unit of oxygen is evolved from the platinum wire in the other. The platinum wire is thus thrown into a condition analogous to zinc, or has a power given to it of determining the oxygen of the liquid to its surface, though it cannot, as is the case with zinc, combine with it under similar circumstances. If we now substitute for the platinum wire, which was connected with the platinum plate, a zinc wire, we have, in addition to the determining tendency by which the platinum was affected, the chemical affinity of the oxygen in vessel B for the zinc wire: thus we have, added to the force which was originally produced by the zinc of the combination in vessel A, a second force produced by the zinc in vessel B, co-operating with the first; two pairs of zinc and platinum thus connected produce, therefore, a more intense effect than one pair; and if we go on adding to these alternations of zinc, platinum, and liquid, we obtain an indefinite exaltation of chemical power, just as in mechanics we obtain accelerated motion by adding fresh impulses to motion already generated.

The same rule of proportion which holds good in chemical combinations also obtains in electrical effects, when these are produced by chemical actions. Dalton and others proved that the constituents of a vast number of compound substances always bore a definite quantitative relation to each other: thus, water, which consists of one part by weight of hydrogen united to eight parts of oxygen, cannot be formed by the same elements in any other than these proportions; you can neither add to nor subtract from the normal ratio of the elements, without entirely altering the nature of the compound. Further, if any element be selected as unity, the combining ratios of other elements will bear an invariable quantitative relation to that and to each other: thus, if hydrogen be chosen as 1, oxygen will be 8, chlorine will be 36; that is, oxygen will unite with hydrogen in the proportion of 8 parts by weight to 1, while chlorine will unite with hydrogen in the proportion of 36 to 1, or with oxygen in the proportion of 36 to 8. Numbers expressing their combining weights, which are thus relative, not absolute, may, by a conventional assent as to the point of unity, be fixed for all chemical reagents; and, when so fixed, it will be found that bodies, at least in inorganic compounds, generally unite in those proportions, or in simple multiples of them: these proportions are termed *Equivalents*.

Now a voltaic battery, which consists usually of alternations of two metals, and a liquid capable of acting chemically upon one of them, has, as we have seen, the power of producing chemical action in a liquid connected with it by metals upon which this liquid is incapable of acting: in such case the constituents of the liquid will be eliminated at the surfaces of the immersed metals, and at a distance one from the other. For example, if the two platinum terminals of a voltaic battery be immersed in water, oxygen will be evolved at one and hydrogen at the other terminal, exactly in the proportions in which they form water; while, to the most minute examination, no action is perceptible in the intervening stratum of liquid. It was known before Faraday's time that, while this chemical action was going on in the subjected liquid, a chemical action was going on in the cells of the voltaic battery; but it was scarcely if at all known that the amount of chemical action in the one bore a constant relation to the amount of action in the other. Faraday proved that it bore a direct equivalent relation: that is, supposing the battery to be formed of zinc, platinum, and water, the amount of oxygen which united with the zinc in each cell of the battery was exactly equal to the amount evolved at the one platinum terminal, while the hydrogen evolved from each platinum plate of the battery was equal to the hydrogen evolved from the other platinum terminal.

Supposing the battery to be charged with hydrochloric acid, instead of water, while the terminals are separated by water, then for every 36 parts by weight of chlorine which united with each plate of zinc, eight parts of oxygen would be evolved from one of the platinum terminals: that is, the weights would be precisely in the same relation which Dalton proved to exist in their chemical combining weights. This relation applies to all liquids capable of being decomposed by the voltaic electrical force, thence called *Electrolytes*: and as no voltaic effect is produced by liquids incapable of being thus decomposed, it follows that voltaic action is chemical action taking place at a distance, or transferred through a chain of media, and that the chemical equivalent numbers are the exponents of the amount of voltaic action for corresponding chemical substances.

As heat, light, magnetism, or motion, can be produced by the requisite application of the electric current, and as this is definitely produced by chemical action, we get these forces very definitely, though not immediately, produced by chemical action. Let us, however, here enquire, as we have already done with respect to the other forces, how far other forces may directly emanate from chemical affinity.

Heat is an immediate product of chemical affinity. I know of no exception to the general proposition that all bodies in chemically combining produce heat; i. e. if solution be not considered as chemical action, and even in that case, when cold results, it is from a change of consistence, as from the solid to the liquid state, and not from chemical action.

We shall find that the same view of the expenditure of force which we have considered in treating of latent heat holds good as to the expenditure of chemical force when regarded with reference to the amount of heat or repulsive force which it engenders, the chemical force being here exhausted by mechanical expansion—that is, by heat. Thus, in the chemical action of the ordinary combustion of coal and oxygen, the expenditure of fuel will be in proportion to the expansibility of the substances heated; water passing freely into the state of steam will consume more fuel than if it be confined and kept at a temperature above its boiling point.

Why chemical action produces heat, or what is the action of the molecules of matter when chemically uniting, is a question upon which many theories have been proposed and which may possibly be never more than approximately resolved.

Some authors explain it by the condensation which takes place; but this will not account for the many instances where, from the liberation of gases, a great increase of volume ensues upon chemical combustion, as in the familiar instance of the explosion of gunpowder: others explain it as resulting from the union of atmospheres of positive and negative electricity which are assumed to surround the atoms of bodies; but this involves hypothesis upon hypothesis. Dr. Wood has lately thrown out the view of the heat of chemical action which is more in accordance with a dynamic theory of heat, and as such demands some notice. Starting with his proposition, which I have previously mentioned, 'that the nearer the particles of bodies are to each other the less they require to move to produce a given motion in the particles of another body,' his argument assumes something of this form.

In the mechanical approximation of the particles of a homogeneous body heat results; the particles $a\ a$ of the body A would, by their approximation, produce expansion in the neighbouring body B, the more so in proportion as they themselves were previously nearer to each other. In chemically combining, $a\ a$ the particles of A are brought into very close proximity with $b\ b$ the particles of B; heat should therefore result, and the greater because the proximity may fairly be assumed to be greater in the case of chemical combination than in that of mechanical compression. In cases, then, where there is no absolute diminution of bulk ensuing on chemical combination, if the greater proximity of the combining particles be such that the correlative expansion ought to be greater (if there were no chemical combination) than that occupied by the total volume of the new compound, an extra expanding power is evolved, and heat or expansion ought to be produced in surrounding bodies. In other words, if *a a* could be brought by physical attraction as near each other as they are by chemical attraction brought near to *b b*, they would, from their increased proximity, produce an expansive power *ultra* the volume occupied by the actual chemical compound A and B. The question, however, immediately occurs, why should the volume of the compound be limited and not occupy the full space equivalent to the expanding power induced by the contraction or approximation of the particles. As the distance of the particles is the resultant of the contending contracting and expanding powers, this result ought to express itself in terms of the actual volume produced by the combination, which it certainly does not.

Though I see some difficulties in Dr. Wood's theory, and perhaps have not rightly conceived it, his views have to my mind great interest, his mode of regarding natural phenomena being analogous to that which I have in this Essay, and for many years, advocated, viz. to divest physical science as much as possible of hypothetic fluids, ethers, latent entities, occult qualities, &c. My own notion of the heat produced by chemical combination, though I scarcely dare venture an opinion upon a subject so controverted, is, that it is analogous to the heat of friction, that the particles of matter in close approximation and rapid motion *inter se* evolve heat as a continuation of the motion interrupted by the friction or intestinal motion of the particles: heat would thus be produced, whether the resulting compound were of greater or less bulk than the sum of the components, though of course when the compound is of greater bulk less heat would be apparent in neighbouring bodies, the expansion taking place in one of the substances themselves—I say in one of them, for it is stated in books of authority that there is no instance of two or more solids or liquids, or a solid and a liquid, combining and producing a compound which is entirely gaseous at ordinary temperatures and pressures. The substance gun-cotton, however, discovered by Dr. Schoenbein, very nearly realises this proposition.

Dr. Andrews has arrived at the conclusion, after careful experiment, that in chemical combinations where acids and alkalies or analogous substances are employed, the amount of heat produced is determined by the basic ingredient, and his experiments have received general assent; although it should be stated that M. Hess arrived at contrary results, the acid constituent according to his experiments furnishing the measure of the heat developed.

Light is directly produced by chemical action, as in the flash of gunpowder, the burning of phosphorus in oxygen gas, and all rapid combustions: indeed, wherever intense heat is developed, light accompanies it. In many cases of slow combustion, such as the phenomena of phosphorescence, the light is apparently much more intense than the heat; the former being obvious, the latter so difficult of detection that for a long time it was a question whether any heat was eliminated; and I am not aware that, at the present day, any thermic effects from certain modes of phosphorescence, such as those of phosphorescent wood, putrescent fish, &c., have been detected.

Chemical action produces *magnetism* whenever it is thrown into a definite direction, as in the phenomenon of electrolysis. I may adduce the gas voltaic battery, as presenting a simple instance of the direct production of magnetism by chemical synthesis. Oxygen and hydrogen in that combination chemically unite; but instead of combining by intimate molecular admixture, as in the ordinary cases, they act upon water, i. e. combined oxygen and hydrogen, placed between them so as to produce a line of chemical action; and a magnet adjacent to this line of action is deflected, and places itself at right angles to it. What a chain of molecules does here, there can be no doubt, all the molecules entering into combination would produce in ordinary chemical actions; but in such cases, the direction of the lines of combination being irregular and confused, there is no general resultant by which the magnet can be affected.

What the exact nature of the transference of chemical power across an electrolyte is, we at present know not, nor can we form any more definite idea of it than that given by the theory of Grotthus. We have no knowledge as to the exact nature of any mode of chemical action, and, for the present, must leave it as an obscure action of force, of which future researches may simplify our apprehension.

We have seen that an equivalent or proportionate electrical effect is produced by a given amount of chemical action; if we, in turn, produce heat, magnetism and motion by the electricity resulting from chemical action, we shall be able to measure these forces far more accurately than when they are directly produced, and thus to deduce their equivalent relation to the initial chemical action. Thus M. Favre, after ascertaining the quantity of heat produced by the oxidation of a given quantity of zinc, and finding, as have others, that the heat is the same when evolved from a voltaic battery by the same consumption of zinc forming its positive element, makes the following experiment.

A voltaic battery and electro-magnet are immersed in calorimeters, and the heat produced when the connection with the magnet is effected is noted.

The electro-magnet is then made to raise a weight, and thus perform mechanical work; and the heat produced is again noted. It is found in the latter case that less heat is evolved than in the former, a certain quantity of heat has therefore been replaced by the mechanical work; and by estimating the amount of heat subtracted, and the amount of work produced, he deduces the relative equivalent of work to heat. These experiments give a production of mechanical work by chemical action, not, it is true, a direct production, but, as the heat and work are in inverse ratios, and each has its source in chemical action, they prove that they are definite for a definite amount of chemical action, and as each is produced respectively by electricity and magnetism, these forces must also bear a definite relation to the initial chemical force.

The doctrine of definite combining proportions, which so beautifully serves to relate chemistry to voltaic electricity, led to the atomic theory, which, though adopted in its universality by a large majority of chemists, presents great difficulties when extended to all chemical combinations.

The equivalent ratios in which a great number of substances chemically combine, hold good in so many instances, that the atomic doctrine is believed by many to be universally applicable, and called a law; and yet, when followed in the combinations of substances whose mutual chemical attractions are very feeble, the relation fades away, and is sought to be recovered by applying a separate and arbitrary multiplier to the different constituents.

Thus, when it was found that a vast number of substances combined in definite volumes and weights, and in definite volumes and weights only, it was argued that their ultimate molecules or atoms had a definite size, as otherwise there was no apparent reason why this equivalent ratio should hold good? why, for instance, water should only be formed of two volumes or one unit by weight of hydrogen, and of one volume or eight units by weight of oxygen? why, unless there were some ultimate limits to the divisibility of its molecules, should not water, or a fluid substance approximating to water in character, be formed by a half, a third, or a tenth part of hydrogen, with eight parts of oxygen?

It was perfectly consistent with the atomic view that a substance might be formed with one part combined with eight parts, or with sixteen, or with twenty-four, for in such a substance there would be no subdivision of the (supposed indivisible) molecule; and this held good with many compounds: thus fourteen parts by weight, say grains of nitrogen, will combine respectively with eight, sixteen, twenty-four, thirty-two, and forty parts by weight, or grains, of oxygen.

So, again, twenty-seven grains of iron will combine with eight grains of oxygen or with twenty-four grains, i. e. three proportionals of oxygen. No compound is known in which twenty-seven grains of iron will combine with two proportionals or sixteen grains of oxygen; but this does not much affect the theory, as such a compound may be yet discovered, or there may be reasons at present unknown why it cannot be formed.

But now comes a difficulty: twenty-seven parts by weight of iron *will* combine with twelve parts by weight of oxygen, and twenty-seven parts of iron will also combine with ten and two-third parts of oxygen. Thus, if we retain the unit of iron we must subdivide the unit of oxygen, or if we retain the unit of oxygen we must subdivide the unit of iron, or we must subdivide both by a different divisor. What then becomes of the notion of an atom or molecule physically indivisible?

If iron were the only substance to which this difficulty applied, it might be viewed as an unexplained exception, or as a mixture of two oxides; or recourse might be had to a more minute subdivision to form the units or equivalents of other substances; but numerous other substances fall under a similar category; and in organic combinations, to preserve the atomic nomenclature, we must apply a separate multiplier or divisor to far the greater number of the elementary constituents, i.e. we must divide that which is, *ex hypothesi*, indivisible.

Thus, to take a more complex substance than any formed by the combination of iron and oxygen, let us select the substance albumen, composed of carbon, hydrogen, nitrogen, oxygen, phosphorus, and sulphur. In this case we must either divide the atoms of phosphorus and sulphur so as to reduce them to small fractions, or multiply the atoms of the other substances by extravagant numbers; thus to preserve the unit of one of the constituents of this substance, chemists say it is composed of 400 atoms of carbon, 310 of hydrogen, 120 of oxygen, 50 of nitrogen, 2 of sulphur, and 1 of phosphorus. This is a somewhat extreme case, but similar difficulties will be found in different degrees to prevail among organic compounds; in very many no constituent can be taken as a unit to which simple multiples of any of the others will give their relative proportions. By the mode of notation adopted, if any conceivable substance be selected, it could, whatever be the proportions of its constituents, be termed atomic. A solution of an ounce of sugar in a pound of water, in a pound and a half, in a pound and a quarter, in a pound and a tenth, might be expressed in an atomic form, if we select arbitrarily a multiplier or divisor.

It is true that in the case of solution, different proportions can be united up to the point of saturation without any difference in the character of the compound, though the same may be predicated to some extent of an acid and an alkali; but even where the steps are sudden, and compounds only exist with definite proportions, they cannot, in a multitude of cases, be reconciled with the true idea of an atomic combination, i.e. one to one, one to two, one to three, &c.

Although, therefore, nature presents us with facts which show that there is some restrictive law of combination which in numerous cases limits the ratios in which substances will combine, nay, further, shows many instances of a proportion between the combining weights of one compound and those of another; although she shows also a remarkable simplicity in the combining volumes of numerous gases, she also gives numerous cases to which the doctrine of atomic combinations cannot fairly be applied.

That there must be something in the constitution of matter, or in the forces which act on it, to account for the *per saltum* manner in which chemical combinations take place, is inevitable; but the idea of atoms does not seem satisfactorily to account for it.

By selecting a separate multiplier or divisor, chemists may denote every combination in terms derived from the atomic theory; but they have passed from the original law, which contemplated only definite multiples, and the very hypothetic expressions of atoms, which the apparently simple relations of combining weights first led them to adopt, they are obliged to vary and to contradict in terms, by dividing that which their hypothesis and the expression of it assumed to be indivisible.

While, therefore, I fully recognise a great natural truth in the definite ratios presented by a vast number of chemical combinations, and in the *per saltum* steps in which nearly all take place, I cannot accept as an argument in favour of an atomic theory, those combinations which are made to support it by the application of an arbitrary notation.

A similar straining of theory seems gradually obtaining in regard to the doctrine of compound radicals. The discovery of cyanogen by Gay-Lussac was probably the first inducement to the doctrine of compound radicals: a doctrine which is now generally, perhaps too generally, received in organic chemistry. As, in the case of cyanogen, a body obviously compound discharged in almost all its reactions the functions of an element, so in many other cases it was found that compound bodies, in which a number of elements existed, might be regarded as binary combinations, by considering certain groups of these elements as a compound radical; that is, as a simple body when treated of in relation to the other complex substances of which it forms part, and only as non-elementary when referred to its internal constitution.

Undoubtedly, by approximating in theory the reactions of inorganic and of organic chemistry, by keeping the mind within the limits of a beaten path, instead of allowing it to wander through a maze of isolated facts, the doctrine of compound radicals has been of service; but, on the other hand, the indefinite variety of changes which may be rung upon the composition of an organic substance, by different associations of its primary elements, makes the binary constituents vary as the minds of the authors who treat of them, and makes their grouping depend entirely upon the strength of the analogies presented to each individual mind. From this cause, and from the extreme license which has been taken in theoretic groupings deduced from this doctrine, a serious question arises whether it may not ultimately, unless carefully restricted, produce confusion rather than simplicity, and be to the student an embarrassment rather than an assistance.

OTHER MODES OF FORCE.

CATALYSIS, or the chemical action induced by the mere presence of a foreign body, embraces a class of facts which must considerably modify many of our notions of chemical action: thus oxygen and hydrogen, when mixed in a gaseous state, will remain unaltered for an indefinite period; but the introduction to them of a slip of clean platinum will cause more or less rapid combination, without being in itself in any respect altered. On the other hand, oxygenated water, which is a compound of one equivalent of hydrogen plus two of oxygen, will, when under a certain temperature, remain perfectly stable; but touch it with platinum in a state of minute division, and it is instantly decomposed, one equivalent of oxygen being set free. Here, again, the platinum is unaltered, and thus we have synthesis and analysis effected apparently by the mere contact of a foreign body. It is not improbable that the increased electrolytic power of water by the addition of some acids, such as the sulphuric and phosphoric, where the acids themselves are not decomposed, depends upon a catalytic effect of these acids;

but we know too little of the nature and rationale of catalysis to express any confident opinion on its modes of action, and possibly we may comprehend very different molecular actions under one and the same name. In no case does catalysis yield us new power or force: it only determines or facilitates the action of chemical force, and, therefore, is no creation of force by contact.

The force so developed by catalysis may be converted into a voltaic form thus: in a single pair of the gas battery previously alluded to, one portion of a strip of platinum is immersed in a tube of oxygen, the other in one of hydrogen, both the gases and the extremities of the platinum being connected by water or other electrolyte; a voltaic combination is thus formed, and electricity, heat, light, magnetism, and motion, produced at the will of the experimenter.

In this combination we have a striking instance of correlative expansions and contractions, analogous, though in a much more refined form, to the expansions and contractions by heat and cold detailed in the early part of this essay, and illustrated by the alternate actions of two bladders partially filled with air: thus, as by the effect of chemical combination in each pair of tubes of the gas battery the gases oxygen and hydrogen lose their gaseous character and shrink into water, so at the platinum terminals of the battery, when immersed in water,

water is decomposed, and expands into oxygen and hydrogen gases. The correlate of the force which changes gas into liquid at one point of space changes liquid into gas at another, and the exact volume which disappears in the one place reappears in the other; so that it would appear to an inexperienced eye as though the gases passed through solid wires.

Gravitation, inertia, and aggregation, were but cursorily alluded to in my original lectures; their relation to the other modes of force seemed to be less definitely traceable; but the phenomenal effects of gravitation and inertia, being motion and resistance to motion, in considering motion I have in some degree included their relations to the other forces.

To my mind gravitation would only produce other force when the motion caused by it is diminished or ceases. Thus, if we suppose a meteor to be a mass rotating in an orbit round the earth, and with no resisting medium, then, as long as that rotation continues, the motion of the meteoric mass itself would be the exponent of the force impelling it; if there be a resisting medium, part of this motion would be arrested and taken up by the medium, either as motion, heat, electricity, or some other mode of force; if the meteor approach the earth sufficiently to fall upon it, the perceptible motion of the meteor is stopped, but is taken up by the earth which vibrates through its mass; part also reappears as heat in both earth and meteor, and part in the change in the earth's position consequent on its increase of gravity, and so on. Gravitation is but the subjective idea, and its relation to other modes of force seems to me to be identical with that of pressure or motion. Thus, when arrested motion produces heat, it matters not whether the motion has been produced by a falling body, i. e. by gravitation, or a body projected by an explosive compound, &c.; the heat will be the same, provided the mass and velocity at the time of arrest be the same. In no other sense can I conceive a relation between gravitation and the other forces, and, with all diffidence, I cannot agree with those who consider there is a different sort of link.

Mosotti has mathematically treated of the identity of gravitation with *cohesive attraction*, and Plücker has recently succeeded in showing that crystalline bodies are definitely affected by magnetism, and take a position in relation to the lines of magnetic force dependent upon their optical axis or axis of symmetry.

What is termed the optic axis is a fixed direction through crystals, in which they do not doubly refract light, and which direction, in those crystals which have one axis of figure, or a line around which the figure is symmetrical, is parallel to the axis of symmetry. When submitted to magnetic influence such crystals take up a position, so that their optic axis points diamagnetically or transversely to the lines of magnetic force; and when, as is the case in some crystals, there is more than one optic axis, the resultant of these axes points diamagnetically. The mineral cyanite is influenced by magnetism in so marked a manner that when suspended it will arrange itself definitely with reference to the direction of terrestrial magnetism, and may, according to Plücker, be used as a compass-needle.

There is scarcely any doubt that the force which is concerned in *aggregation* is the same which gives to matter its crystalline form; indeed, a vast number of inorganic bodies, if not all, which appear amorphous, are, when closely examined, found to be crystalline in their structure: we thus get a reciprocity of action between the force which unites the molecules of matter and the magnetic force, and through the medium of the latter the correlation of the attraction of aggregation with the other modes of force may be established.

I believe that the same principles and mode of reasoning as have been adopted in this essay might be applied to the organic as well as the inorganic world; and that muscular force, animal and vegetable heat, &c., might, and at some time will, be shown to have similar definite correlations; but I have purposely avoided this subject, as pertaining to a department of science to which I have not devoted my attention. I ought, however, while alluding to this subject, shortly to mention some experiments of Professor Matteucci, communicated to the Royal Society in the year 1850, by which it appears that whatever mode of force it be which is propagated along the nervous filaments, this mode of force is definitely affected by currents of electricity. His experiments show that when a current of positive electricity traverses a portion of the muscle of a living animal in the same direction as that in which the nerves ramify—i.e. a direction from the brain to the extremities—a muscular contraction is produced in the limb experimented on, showing that the nerve of motion is affected; while, if the current, as it is termed, be made to traverse the muscle in the reverse direction, or towards the nervous centres, the animal utters cries, and exhibits all the indications of suffering pain, scarcely any muscular movement being produced; showing that in this case the nerves of sensation are affected by the electric current, and therefore that some definite polar condition exists, or is induced, in the nerves, to which electricity is correlated, and that probably this polar condition constitutes or conveys nervous agency. There are other analogies given in the papers of M. Matteucci, and derived from the action of the electrical organs of fishes, which tend to corroborate and develope the same view.

By an application of the doctrine of the Correlation of Forces, Dr. Carpenter has shown how a difficulty arising from the ordinary notions of the development of an organised being from its germ-cell may be lessened. It has been thought by many physiologists that the *nisus formativus*, or organising force of an animal or vegetable structure, lies dormant in the primordial germ-cell. 'So that the organising force required to build up an oak or a palm, an elephant or a whale, is concentrated in a minute particle only discernible by microscopic aid.'

Certain other views of nearly equal difficulty have been propounded. Dr. Carpenter suggests the probability of extraneous forces, as heat, light, and chemical affinity, continuously operating upon the material germ; so that all that is required in this is a structure capable of receiving, directing, and converting these forces into those which tend to the assimilation of extraneous matter and the definite development of the particular structure. In proof of this position he shows how dependent the process of germ development is upon the presence and agency of external forces, particularly heat and light, and how it is regulated by the measure of these forces supplied to it.

It certainly is far less difficult so to conceive the

supply of force yielded to organised beings in their gradual process of growth, than to suppose a store of dormant or latent force pent up in a microscopic monad.

As by the artificial structure of a voltaic battery, chemical actions may be made to co-operate in a definite direction, so, by the organism of a vegetable or animal, the mode of motion which constitutes heat, light, &c., may, without extravagance, be conceived to be appropriated and changed into the forces which induce the absorption, and assimilation of nutriment, and into nervous agency and muscular power. Indications of similar thoughts may be detected in the writings of Liebig.

Some difficulty in studying the correlations of vital with inorganic physical forces arises from the effects of sensation and consciousness, presenting a similar confusion to that alluded to, when, in treating of heat, I ventured to suggest, that observers are too apt to confound the sensations with the phenomena. Let us apply some of the considerations on force, given in the introductory portion of this essay, to cases where vitality or consciousness intervenes. When a weight is raised by the hand, there should, according to the doctrine of non-creation of force, have been somewhere an expenditure equivalent to the amount of gravitation overcome in raising the weight. That there is expenditure we can prove, though in the present state of science we cannot measure it. Thus, prolong the effort, raise weights for an hour or two, the vital powers sink, food, i. e. fresh chemical force, is required to supply the exhaustion. If this supply is withheld and the exertion is continued, we see the consumption of force in the supervening weakness and emaciation of the body.

The consciousness of effort, which has formed a topic of argument by some writers when treating of force, and is by them believed to be that which has originated the idea of force, may by the physical student be regarded as feeling is in the phenomena of heat and cold, viz. a sensation of the struggle of opposing molecular motions in overcoming the resistance of the masses to be moved. When we say we feel hot, we feel cold, we feel that we are exerting ourselves, our expressions are intelligible to beings who are capable of experiencing similar sensations; but the physical changes accompanying these sensations are not thereby explained. Without pretending to know what probably we shall never know, the actual *modus agendi* of the brain, nerves, muscles, &c., we may study vital as we do inorganic phenomena, both by observation and experiment. Thus, Sir Benjamin Brodie has examined the effect of respiration on animal heat by inducing artificial respiration after the spinal cord has been severed; in which case he finds the animal heat declines, notwithstanding the continuance of the chemical action of respiration, carbonic acid being formed as usual; but he also finds that under such circumstances the struggles or muscular actions of the animal are very great, and sufficient probably to account for the force eliminated by the chemical action in digestion and respiration; and Liebig, by measuring the amount of chemical action in digestion and respiration, and comparing it with the labour performed, has to some extent established their equivalent relations.

Mr. Helmholtz has found that the chemical changes which take place in muscles are greater when these are made to undergo contractions than when they are in repose; and that, as would be expected, the consumption of the matter of the muscle, or, in other terms, the waste of excrementitious matter thrown off, is greater in the former than in the latter case.

M. Matteucci has ascertained that the muscles of recently killed frogs absorb oxygen and exhale carbonic acid, and that when they are thrown into a state of contraction, and still more when they perform mechanical work, the absorption is increased; and he even calculates the equivalents of work so performed.

M. Beclard finds that the quantity of heat

produced by voluntary muscular contraction in man is greater when that contraction is what he terms static, that is, when it produces no external work, but is effort alone, than when that effort and contraction are employed dynamically, so as to raise a weight or produce mechanical work.

Thus, though we may see no present promise of being able to resolve sensations into their ultimate elements, or to trace, physically, the link which unites volition with exertion or effort, we may hope to approximate the solution of these deeply interesting questions.

In the same individual the chemical and physical state of the secretions in the warm may be compared with those in the cold parts of the body. The changes in digestion and respiration, when the body is in a state of rest, may be compared with those which obtain when it is in a state of activity. The relations with external matter, maintaining by the constant play of natural forces, the vital nucleus, or the organisation by means of which matter and force receive, for a definite period, a definite incorporation and direction, may be ascertained, while the more minute structural changes are revealed to us by the ever improving powers of the microscope; and thus step by step we may learn that which it is given to us to learn, boundless in its range and infinite in its progress, and therefore never giving a response to the ultimate—Why?

As the first glimpse of a new star is caught by the eye of the astronomer while directing his vision to a different point of space, and disappears when steadfastly gazed at, only to have its position and figure ultimately ascertained by the employment of more penetrative powers, so the first scintillations of new natural phenomena frequently present themselves to the eye of the observer, dimly seen when viewed askance, and disappearing if directly looked for. When new powers of thought and experiment have developed and corrected the first notions, and given a character to the new image, probably very different from the first impression, fresh objects are again glanced at in the margin of the new field of vision, which in their turn have to be verified, and again lead to new extensions; thus the effort to establish one observation leads to the imperfect perception of new and wider fields of research; and, instead of approaching finality, the more we discover the more infinite appears the range of the undiscovered!

CONCLUDING REMARKS.

I HAVE now gone through the affections of matter to which distinct names have been given in our received nomenclature: that other forces may be detected, differing as much from them as they differ from each other, is highly probable, and that when discovered, and their modes of action fully traced out, they will be found to be related *inter se*, and to these forces as these are to each other, I believe to be as far certain as certainty can be predicted of any future event.

It may in many cases be a difficult question to determine what constitutes a distinct affection of matter or mode of force. It is highly probable that different lines of demarcation would have been drawn between the forces already known, had they been discovered in a different manner, or first observed at different points of the chain which connects them. Thus, radiant heat and light are mainly distinguished by the manner in which they affect our senses: were they viewed according to the way in which they affect inorganic matter, very different notions would possibly be entertained of their character and relation. Electricity, again, was named from the substance in which, and magnetism from the district where, it first happened to be observed, and a chain of intermediate phenomena have so connected electricity with galvanism that they are now regarded as the same force, differing only in the degree of its intensity and quantity, though for a long time they were regarded as distinct.

The phenomenon of attraction and repulsion by amber, which originated the term *electricity*, is as unlike that of the decomposition of water by the voltaic pile, as any two natural phenomena can well be. It is only because the historical sequence of scientific discoveries has associated them by a number of intermediate links, that they are classed under the same category. What is called voltaic electricity might equally, perhaps more appropriately, be called volta'c chemistry. I mention these facts to show that the distinction in the name may frequently be much greater than the distinction of the subject which it represents, and vice versâ, not as at all objecting to the received nomenclature on these points; nor do I say it would be advisable to depart from it : were we to do so, inevitable confusion would result, and objections equally forcible might be found to apply to our new terminology.

Words, when established to a certain point, become a part of the social mind; its powers and very existence depend upon the adoption of conventional symbols; and were these suddenly departed from, or varied, according to individual apprehensions, the acquisition and transmission of knowledge would cease. Undoubtedly, neology is more permissible in physical science than in any other branch of knowledge, because it is more progressive; new facts or new relations require new names, but even here it should be used with great caution.

Si forte necesse est
Indiciis monstrare recentibus abdita rerum,
Fingere cinctutis non exaudita Cethegis
Continget; dabiturque licentia, sumpta pudenter.

Even should the mind ever be led to dismiss the idea of various forces, and regard them as the exertion of one force, or resolve them definitely into motion; still we could never avoid the use of different conventional terms for the different modes of action of this one pervading force.

Reviewing the series of relations between the various forces which we have been considering, it would appear that in many cases where one of these is excited or exists, all the others are also set in action: thus, when a substance, such as sulphuret of antimony, is electrified, at the instant of electrisation it becomes *magnetic* in directions at right angles to the lines of electric force; at the same time it becomes *heated* to an extent greater or less according to the intensity of the electric force. If this intensity be exalted to a certain point the sulphuret becomes luminous, or *light* is produced: it expands, consequently *motion* is produced; and it is decomposed, therefore *chemical action* is produced. If we take another substance, say a metal, all these forces except the last are developed; and although we can scarcely apply the term chemical action to a substance hitherto undecomposed, and which, under the circumstances we are considering, enters into no new combination, yet a metal undergoes that species of polarisation which, as far as we can judge, is the first step towards chemical action, and which, if the substance were decomposable, would resolve it into its elements. Perhaps, indeed, some hitherto undiscovered chemical action is produced in substances which we regard as undecomposable: there are experiments to show that metals which have been electrised are permanently changed in their molecular constitution. Oxygen, we have seen, is changed by the electric spark into ozone, and phosphorus into allotropic phosphorus, both which changes were for a long time unknown to those familiar with electrical science.

Thus, with some substances, when one mode of force is produced all the others are simultaneously developed. With other substances, probably with all matter, some of the other forces are developed, whenever one is excited, and all may be so were the matter in a suitable condition for their developement, or our means of detecting them sufficiently delicate.

This simultaneous production of several different forces seems at first sight to be irreconcileable with their mutual and necessary dependence, and it certainly presents a formidable experimental difficulty in the way of establishing their equivalent relations; but when examined closely, it is not in fact inconsistent with the views we have been considering, but is indeed a strong argument in favour of the theory which regards them as modes of motion.

Let us select one or two cases in which this form of objection may be prominently put forward. A voltaic battery decomposing water in a voltameter, while the same current is employed at the same time to make an electro-magnet, gives nevertheless in the voltameter an equivalent of gas, or decomposes an equivalent of an electrolyte for each equivalent of chemical decomposition in the battery cells, and will give the same ratios if the electro-magnet be removed. Here, at first sight, it would appear that the magnetism was an extra force produced, and that thus more than the equivalent power was obtained from the battery. In answer to this objection it may be said, that in the circumstances under which this experiment is ordinarily performed, several cells of the battery are used, and so there is a far greater amount of force generated in the cells than is indicated by the effect in the voltameter. If, moreover, the magnet be not interposed, still the magnetic force is equally existent throughout the whole current; for instance, the wires joining the plates will attract iron filings, deflect magnetic needles, &c., and produce diamagnetic effects on surrounding matter. By the iron core a small portion of the force is, indeed, absorbed *while* it is being made a magnet, but this ceases to be absorbed when the magnet is made; this has been proved by the observation of Mr. Latimer Clarke, who has found that along the wires of the electric telegraph the magnetic needles placed at different stations remained fixed after the connection with the battery was made, and while the electric current acted by induction on surrounding conducting matter, separated from the wires by their gutta percha coating, so that a sort of Leyden phial was formed; but as soon as this induction had produced its effect between each station, or, so to speak, the phial was charged, the needles successively were deflected: it is like the case of a pulley and weight, which latter exhausts force while it is being raised; but when raised, the force is free, and may be used for other purposes.

If a battery of one cell, just capable of decomposing water and no more, be employed, this will cease to decompose while making a magnet. There must, in every case, be preponderating chemical affinity in the battery cells, either by the nature of its elements or by the reduplication of series, to effect decomposition in the voltameter; and if the point is just reached at which this is effected, and the power is then reduced by any resistance, decomposition ceases: were it otherwise, were the decomposition in the voltameter the exponent of the entire force of the generating cells, and these could independently produce magnetic force, this latter force would be got from nothing, and perpetual motion be obtained.

To take another and different example: A piece of zinc dissolved in dilute sulphuric acid gives somewhat less heat than when the zinc has a wire of platinum attached to it, and is dissolved by the same quantity of acid. The argument is deducible that, as there is more electricity in the second than in the first case, there should be less heat; but as, according to our received theories, the heat is a product of the electric current, and in consequence of the impurity of zinc electricity is generated in the first case molecularly, by what is called local action, though not thrown into a general direction, there should be more of both heat and electricity in the second than in the first case, as the heat and electricity due to the voltaic combination of zinc and platinum are added to that excited on the surface of the zinc, and the zinc should be, as in fact it is, more rapidly dissolved; so that the extra heat and electricity are produced by extra chemical force. Many additional cases of a similar description might be suggested. But although it is difficult, and perhaps impossible, to restrict the action of any one force to the production of one other force, and of one only — yet if the whole of one force, say chemical action, be supposed to be employed in producing its full equivalent of another force, say heat, then as this heat is capable in its turn of reproducing chemical action, and in the limit, a quantity equal or at least only infinitely short of the initial force: if this could at the same time produce independently another force, say magnetism, we could, by adding the magnetism to the total heat, get more than the original chemical action, and thus create force or obtain perpetual motion.

The term Correlation, which I selected as the title of my Lectures in 1843, strictly interpreted, means a necessary mutual or reciprocal dependence of two ideas, inseparable even in mental conception: thus, the idea of height cannot exist without involving the idea of its correlate, depth; the idea of parent cannot exist without involving the idea of offspring. It has been scarcely, if at all, used by writers on physics, but there are a vast variety of physical relations to which, if it does not in its strictest original sense apply, cannot certainly be so well expressed by any other term. There are, for example, many facts, one of which cannot take place without involving the other; one arm of a lever cannot be depressed without the other being elevated—the finger cannot press the table without the table pressing the finger. A body cannot be heated without another being cooled, or some other force being exhausted in an equivalent ratio to the production of heat; a body cannot be positively electrified without some other body being negatively electrified, &c.

The probability is, that, if not all, the greater number of physical phenomena are correlative, and that, without a duality of conception, the mind cannot form an idea of them: thus motion cannot be perceived or probably imagined without parallax or relative change of position. The world was believed fixed, until, by comparison with the celestial bodies, it was found to change its place with regard to them: had there been no perceptible matter external to the world, we should never have discovered its motion. In sailing along a river, the stationary vessels and objects on the banks seem to move past the observer: if at last he arrives at the conviction that he is moving, and not these objects, it is by correcting his senses by reflection derived from a more extensive previous use of them: even then he can only form a notion of the motion of the vessel he is in, by its change of position with regard to the objects it passes—

that is, provided his body partakes of the motion of the vessel, which it only does when its course is perfectly smooth, otherwise the relative change of position of the different parts of the body and the vessel inform him of its alternating, though not of its progressive movement. So in all physical phenomena, the effects produced by motion are all in proportion to the relative motion: thus, whether the rubber of an electrical machine be stationary, and the cylinder mobile, or the rubber mobile and the cylinder stationary, or both mobile in different directions, or in the same direction with different degrees of velocity, the electrical effects are, *cæteris paribus*, precisely the same, provided the relative motion is the same, and so, without exception, of all other phenomena. The question of whether there can be absolute motion, or, indeed, any absolute isolated force, is purely the metaphysical question of idealism or realism—a question for our purpose of little import; sufficient for the purely physical inquirer, the maxim '*de non apparentibus et non existentibus eadem est ratio.*'

The sense I have attached to the word correlation, in treating of physical phenomena, will, I think, be evident from the previous parts of this essay, to be that of a necessary reciprocal production; in other words, that any force capable of producing another may, in its turn, be produced by it—nay, more, can be itself resisted by the force it produces, in proportion to the energy of such production, as action is ever accompanied and resisted by reaction: thus, the action of an electro-magnetic machine is reacted upon by the magneto-electricity developed by its action.

To many, however, of the cases we have been considering, the term correlation may be applied in a more strict accordance with its original sense: thus, with regard to the forces of electricity and magnetism in a dynamic state, we cannot electrise a substance without magnetising it—we cannot magnetise it without electrising it:—each molecule, the instant it is affected by one of these forces, is affected by the other; but, in transverse directions, the forces are inseparable and mutually dependent—correlative, but not identical.

The evolution of one force or mode of force into another has induced many to regard all the different natural agencies as reducible to unity, and as resulting from one force which is the efficient cause of all the others: thus, one author writes to prove that electricity is the cause of every change in matter; another, that chemical action is the cause of everything; another, that heat is the universal cause, and so on. If, as I have stated it, the true expression of the fact is, that each mode of force is capable of producing the others, and that none of them can be produced but by some other as an anterior force, then any view which regards either of them as abstractedly the efficient cause of all the rest, is erroneous: the view has, I believe, arisen from a confusion between the abstract or generalised meaning of the term cause, and its concrete or special sense; the word itself being indiscriminately used in both these senses.

Another confusion of terms has arisen, and has, indeed, much embarrassed me in enunciating the propositions put forth in these pages, on account of the imperfection of scientific language; an imperfection in great measure unavoidable, it is true, but not the less embarrassing. Thus, the words light, heat, electricity, and magnetism, are constantly used in two senses—viz. that of the force producing, or the subjective idea of force or power, and of the effect produced, or the objective phenomenon. The word motion, indeed, is only applied to the effect, and not to the force, and the term chemical affinity is generally applied to the force, and not to the effect; but the other four terms are, for want of a distinct terminology, applied indiscriminately to both.

I may have occasionally used the same word at one time in a subjective, at another in an objective sense; all I can say is, that this cannot be avoided without a neology, which I have not the presumption to introduce, or the authority to enforce. Again, the use of the term forces in the plural might be objected to by those who do not attach to the term force the notion of a specific agency, but of one universal power associated with matter, of which its various phenomena are but diversely modified effects.

Whether the imponderable agents, viewed as force, and not as matter, ought to be regarded as distinct forces or as distinct modes of force, is probably not very material, for, as far as I am aware, the same result would follow either view; I have therefore used the terms indiscriminately, as either happened to be the more expressive for the occasion.

Throughout this essay I have placed motion in the same category as the other affections of matter. The course of reasoning adopted in it, however, appears to me to lead inevitably to the conclusion that these affections of matter are themselves modes of motion; that, as in the case of friction, the gross or palpable motion, which is arrested by the contact of another body, is subdivided into molecular motions or vibrations, which vibrations are heat or electricity as the case may be; so the other affections are only matter moved or molecularly agitated in certain definite directions. I have already considered the hypothesis that the passage of electricity and magnetism causes vibrations in an ether permeating the bodies through which the current is transmitted, or the application of the same ethereal hypothesis to these imponderables which had previously been applied to light; many, in speaking of some of their effects, admit that electricity and magnetism cause or produce by their passage vibrations in the particles of matter, but regard the vibrations produced as an occasional, though not always a necessary, effect of the passage of electricity, or of the increment or decrement of magnetism. The view which I have taken is, that such vibrations, molecular polarisations, or motions of some sort from particle to particle, are themselves electricity or magnetism; or, to express it in the converse, that dynamic electricity and magnetism are themselves motion, and that permanent magnetism, and static electricity, are conditions of force bearing a similar relation to motion which tension or gravitation do.

This theory might well be discussed in greater detail than has been used in this work; but to do this and to anticipate objections would lead into specialities foreign to my present object, in the course of this essay my principal aim having been rather to show the relation of forces as evinced by acknowledged facts, than to enter upon any detailed explanation of their specific modes of action.

Probably man will never know the ultimate structure of matter or the minutiæ of molecular actions; indeed it is scarcely conceivable that the mind can ever attain to this knowledge; the monad irresolvable by a given microscope may be resolved by an increase in power. Much harm has already been done by attempting hypothetically to dissect matter and to discuss the shapes, sizes, and numbers of atoms, and their atmospheres of heat, ether, or electricity.

Whether the regarding electricity, light, magnetism, &c., as simply motions of ordinary matter, be or be not admissible, certain it is, that all past theories have resolved, and all existing theories do resolve, the actions of these forces into motion. Whether it be that, on account of our familiarity with motion, we refer other affections to it, as to a language which is most easily construed and most capable of explaining them, or whether it be that it is in reality the only mode in which our minds, as contradistinguished from our senses, are able to conceive material agencies; certain it is, that since the period at which the mystic notions of spiritual or preternatural powers were applied to account for physical phenomena, all hypotheses framed to explain them have resolved them into motion. Take, for example, the theories of light to which I have before alluded: one of these supposes light to be a highly rare matter, emitted from—i.e. put in *motion* by—luminous bodies; a second supposes that the matter is not emitted from luminous bodies, but that it is put into a state of vibration or undulation, i.e. *motion*, by them; and thirdly, light may be regarded as an undulation or *motion* of ordinary matter, and propagated by undulations of air, glass, &c., as I have before stated. In all these hypotheses, matter and motion are the only conceptions. Nor, even if we accept terms derived from our own sensations, the which sensations themselves may be but modes of motion in the nervous filaments, can we find words to describe phenomena other than those expressive of matter and motion. We in vain struggle to escape from these ideas; if we ever do so, our mental powers must undergo a change of which at present we see no prospect.

If we apply to any other force the mode of reasoning which we have in the course of this essay applied to heat, we shall arrive at the same conclusion, and see that a given source of power can, supposing it to be fully utilised in each case, yield no more by employing it as an exciter of one force than of another. Let us take electricity as an example. Suppose a pound of mercury at 400° be employed to produce a thermo-electric current, and the latter be in its turn employed to produce mechanical force; if this latter force be greater than that which the direct effect of heat would produce, then it could by compression raise the temperature of the mercury itself, or of a similar quantity equally heated, to a higher point than its original temperature, the 400° to 401°, for example, which is obviously impossible; nor, if we admit force to be indestructible, can it produce less than 400°, except by some portion of it being converted into another form or mode of force.

But as the mechanical effect here is produced through the medium of electricity, and the mechanical effect is definite, so the quantity of electricity producing it must be definite also, for unequal quantities of electricity could only produce an equal mechanical effect by a loss or gain of their own force into or out of nothing. The same reasoning will apply to the other forces, and will lead, it appears to me, necessarily and inevitably to the conclusion, that each force is definitely and equivalently convertible into any other, and that where experiment does not give the full equivalent, it is because the initial force has been dissipated, not lost, by conversion into other unrecognised forces. The equivalent is the limit never practically reached.

The great problem which remains to be solved, in regard to the correlation of physical forces, is this establishment of their equivalents of power, or their measurable relation to a given standard. The progress made in some of the branches of this inquiry has been already noticed. Viewed in their static relations, or in the conditions requisite for producing equilibrium or quantitative equality of force, a remarkable relation between chemical affinity and heat is that discovered in many simple bodies by Dulong and Petit, and extended to compounds by Neumann and Avogadro. Their researches have shown that the specific heats of certain substances, when multiplied by their chemical equivalents, give a constant quantity as product—or, in other words, that the combining weights of such substances are those weights which require equal accessions or abstractions of heat, equally to raise or lower their temperature. To put the proposition more in accordance with the view we have taken of the nature of heat: each body has a power of communicating or receiving molecular repulsive power, exactly equal, weight for weight, to its chemical or combining power. For instance, the equivalent of lead is 104, of zinc 33, or, in round numbers, as 3 to 1: these numbers are therefore inversely the exponents of their chemical power, three times as much lead as zinc being required to saturate the same quantity of an acid or substance combining with it; but their power of communicating or abstracting heat or repulsive power is precisely the same, for three times as much lead as zinc is required to produce the same amount of expansion or contraction in a given quantity of a third substance, such as water.

Again, a great number of bodies chemically combine in equal volumes, i.e. in the ratios of their specific gravities; but the specific gravities represent the attractive powers of the substances, or are the numerical exponents of the forces tending to produce motion in masses of matter towards each other; while the chemical equivalents are the exponents of the affinities or tendencies of the molecules of dissimilar substances to combine, and saturate each other; consequently, here we have in certain cases an equivalent relation between these two modes of force — gravitation and chemical attraction.

Were the above relations extended into an universal law, we should have the same numerical expression for the three forces of heat, gravity, and affinity; and as electricity and magnetism are quantitatively related to them, we should have a similar expression for these forces: but at present the bodies in which this parity of force has been discovered, though in themselves numerous, are small compared with the exceptions, and, therefore, this point can only be indicated as promising a generalisation, should subsequent researches alter our knowledge as to the elements and combining equivalents of matter.

With regard to what may be called dynamic equivalents, i.e. the definite relation to time of the action of these varied forces upon equivalents of matter, the difficulty of establishing them is still greater. If the proposition which I stated at the commencement of this paper be correct, that motion may be subdivided or changed in character, so as to become heat, electricity, &c., it ought to follow that when we collect the dissipated and changed forces, and reconvert them, the initial motion, minus an infinitesimal quantity, affecting the same amount of matter with the same velocity, should be reproduced, and so of the changes in matter produced by the other forces; but the difficulties of proving the truth of this by experiment will, in many cases, be all but insuperable; we cannot imprison motion as we can matter, though we may to some extent restrain its direction.

The term perpetual motion, which I have not unfrequently employed in these pages, is itself equivocal. If the doctrines here advanced be founded, all motion is, in one sense, perpetual. In masses whose motion is stopped by mutual concussion, heat or motion of the particles is generated; and thus the motion continues, so that if we could venture to extend such thoughts to the universe, we should assume the same amount of motion affecting the same amount of matter for ever. Where force opposes force, as in cases of static equilibrium, the balance of pre-existing equilibrium is affected, and fresh motion is started equivalent to that which is withdrawn into a state of abeyance.

But the term perpetual motion is applied, in ordinary parlance (and in such sense I have used it), to a perpetual recurrent motion, e.g., a weight which by its fall would turn a wheel, which wheel would, in its turn, raise the initial weight, and so on for ever, or until the material of which the machine is made be worn out. It is strange that to common apprehension the impossibility of this is not self-evident: if the initial weight is to be raised by the force it has itself generated, it must necessarily generate a force greater than that of its own weight or centripetal attraction; in other words, it must be capable of raising a weight heavier than itself: so that, setting aside the resistance of friction, &c., a weight, to produce perpetual recurrent motion, must be heavier than an equal weight of matter, in short, heavier than itself.

Suppose two equal weights at each end of an equi-armed lever, there is no motion; cut off a fraction of one of them, and it rises while the other falls. How, now, is the lesser weight to bring back the greater without any extraneous application of force? If, as is obvious, it cannot do so in this simple form of experiment, it is à fortiori more impossible if machinery be added, for increased resistances have then to be overcome. Can we again mend this by employing any other force? Suppose we employ electricity, the initial weight in descending turns a cylinder against a cushion, and so generates electricity; to make this force recurrent, the electricity so generated must, in its turn, raise the initial weight, or one heavier than it, i.e. the initial weight must, through the medium of the electricity it generates, raise a weight heavier than itself. The same problem, applied to any other forces, will involve the same absurdity: and yet, simple as the matter seems, the world is hardly yet disabused of an idea little removed from superstition.

But the importance of the deductions to be derived from the negation of perpetual motion seems scarcely to have impressed philosophers, and we only find here and there a scattered hint of the consequences necessarily resulting from that which to the thinking mind is a conviction. Some of these I have ventured to put forward in the present essay, but many remain, and will crowd upon the mind of those who pursue the subject. Does not, for instance, the impossibility of perpetual motion, when thought out, involve the demonstration of the impossibility, to which I have previously alluded, of any event identically recurring?

The pendulum in vacuo, at each beat, leaves a portion of the force which started it in the form of heat at its point of suspension; this force, though ever existent, can never be restored in its integrity to the ball of the pendulum, for in the process of restoration it must affect other matter, and alter the condition of the universe. To restore the initial force in its integrity, everything as it existed at the moment of the first beat of the pendulum must be restored in its integrity: but how can this be—for while the force was escaping from the pendulum by radiating heat from the point of suspension, surrounding matter has not stood still; the very attraction which caused the beat of the pendulum has changed in degree, for the pendulum is nearer to or farther from the sun, or from some planet or fixed star.

It might be an interesting and not profitless speculation to follow out these and other consequences; it would, I believe, lead us to the conviction that the universe is ever changing, and that notwithstanding secular recurrences which would primâ facie seem to replace matter in its original position, nothing in fact ever returns or can return to a state of existence identical with a previous state. But the field is too illimitable for me to venture farther.

The inevitable dissipation or throwing off a portion of the initial force presents a great experimental difficulty in the way of establishing the equivalents of the various natural forces. In the steam-engine, for instance, the heat of the furnace not only expands the water and thereby produces the motion of the piston, but it also expands the iron of the boiler of the cylinder and all surrounding bodies. The force expended in expanding this iron to a very small extent is equal to that which expands the vapour to a very large extent: this expansion of the iron is capable, in its turn, of producing a great mechanical force, which is practically lost. Could all the force be applied to the vapour, an enormous addition of power would be gained for the same expenditure; and perhaps even with our present means more might be done in utilising the expansion of the iron.

Another great difficulty in experimentally ascertaining the dynamic equivalents of different forces arises from the effects of disruption, or the overcoming an existing force. Thus, when a part of the initial force employed is engaged in twisting or tearing asunder matter previously held together by cohesive attraction, or in overcoming gravitation or inertia, the same amount of heat or electricity would not apparently be evolved as if such obstacle were non-existent, and the initial force were wholly employed in producing, not in opposing. There is great difficulty in devising experiments in which some portion of the force is not so employed.

The initial force, however, that has been employed for such disruption is not lost, as at the moment of disruption the bodies producing it fly off, and carry with them their force. Thus, let two weights be attached to a cord placed across a bar; when their force is sufficient to break the cord or the bar, the weights fall and strike the earth, making it vibrate, and so conveying away or continuing the force expressed by the cohesion of the bar or cord. If, instead of breaking a cord, the weights be employed to bend a bar, their gravitating force, instead of making the earth vibrate, produces heat in the bar, and so with whatever other force be employed to produce effects of disruption, torsion, &c., so that, though difficult in practice, the numerical problem of the equivalent of the force is not theoretically irresolvable.

The voltaic battery affords us the best means of ascertaining the dynamic equivalents of different forces, and it is probable that by its aid the best theoretical and practical results will be ultimately attained.

In investigating the relation of the different forces, I have in turn taken each one as the initial force or starting-point, and endeavoured to show how the force thus arbitrarily selected could mediately or immediately produce and be merged into the others: but it will be obvious to those who have attentively considered the subject, and brought their minds into a general accordance with the views I have submitted to them, that no force can, strictly speaking, be initial, as there must be some anterior force which produced it: we cannot create force or motion any more than we can create matter.

Thus, to take an example previously noticed, and recede backwards; the spark of light is produced by electricity, electricity by motion, and motion is produced by something else, say a steam-engine—that is, by heat. This heat is produced by chemical affinity, i. e. the affinity of the carbon of the coal for the oxygen of the air: this carbon and this oxygen have been previously eliminated by actions difficult to trace, but of the pre-existence of which we cannot doubt, and in which actions we should find the conjoint and alternating effects of heat, light, chemical affinity, &c. Thus, tracing any force backwards to its antecedents, we are merged in an infinity of changing forms of force; at some point we lose it, not because it has been in fact created at any definite point, but because it resolves itself into so many contributing forces, that the evidence of it is lost to our senses or powers of detection; just as, in following it forward into the effect it produces, it becomes, as I have before stated, so subdivided and dissipated as to be equally lost to our means of detection.

Can we, indeed, suggest a proposition, definitely conceivable by the mind, of force without antecedent force? I cannot, without calling for the interposition of creative power, any more than I can conceive the sudden appearance of a mass of matter come from nowhere, and formed from nothing. The impossibility, humanly speaking, of creating or annihilating matter, has long been admitted, though, perhaps, its distinct reception in philosophy may be set down to the overthrow of the doctrine of Phlogiston, and the reformation of chemistry at the time of Lavoisier. The reasons for the admission of a similar doctrine as to force appear to be equally strong. With regard to matter, there are many cases in which we never practically prove its cessation of existence, yet we do not the less believe in it: who, for instance, can trace, so as to re-weigh, the particles of iron worn off the tire of a carriage wheel? who can recombine the particles of wax dissipated and chemically changed in the burning of a candle? By placing matter undergoing physical or chemical changes under special limiting circumstances, we may, indeed, acquire evidence of its continued existence, weight for weight—and so we may in some instances of force, as in definite electrolysis: indeed, the evidence we acquire of the continued existence of matter is by the continued exertion of the force it exercises, as, when we weigh it, our evidence is the force of attraction; so, again, our evidence of force is the matter it acts upon. Thus, matter and force are correlates, in the strictest sense of the word; the conception of the existence of the one involves the conception of the existence of the other: the quantity of matter again, and the degree of force, involve conceptions of space and time. But to follow out these abstract relations would lead me too far into the alluring paths of metaphysical speculation.

That the theoretical portions of this essay are open to objection I am fully conscious. I cannot, however, but think that the fair way to test a theory is to compare it with other theories, and to see whether upon the whole the balance of probability is in its favour. Were a theory open to no objection it would cease to be a theory, and would become a law; and were we not to theorise, or to take generalised views of natural phenomena until those generalisations were sure and unobjectionable —in other words, were laws—science would be lost in a complex mass of unconnected observations, which would probably never disentangle themselves. Excess on either side is to be avoided; although we may often err on the side of hasty generalisation, we may equally err on the side of mere elaborate collection of observations, which, though sometimes leading to a valuable result, yet, when cumulated without a connecting link, frequently occasion a costly waste of time, and leave the subject to which they refer in greater obscurity than that in which it was involved at their commencement.

Collections of facts differ in importance, as do theories: the former, in many instances, derive their value from their capability of generalisation; while, conversely, theories are valuable as methods of co-ordinating given series of facts, and more valuable in proportion as they require fewer exceptions and fewer postulates. Facts may sometimes be as well explained by one view as by another, but without a theory they are unintelligible and incommunicable. Let us use our utmost effort to communicate a fact without using the language of theory, and we fail; theory is involved in all our expressions; the knowledge of bygone times is imported into succeeding times by terms involving theoretic conceptions. As the knowledge of any particular science developes itself, our views of it become more simple; hypotheses, or the introduction of supposititious views, are more and more dispensed with; words become applicable more directly to the phenomena, and, losing the hypothetic meaning which they necessarily possessed at their reception, acquire a secondary sense, which brings more immediately to our minds the facts of which they are indices. The scaffolding has served its purpose. The hypothesis fades away, and a theory, or generalised view of phenomena, more independent of supposition, but still full of gaps and difficulties, takes its place. This in its turn, should the science continue to progress, either gives place to a more simple and wider generalisation, or becomes, by the removal of objections, established as a law. Even in this more advanced stage words importing theory must be used, but phenomena are now intelligible and connected, though expressed by varied forms of speech.

To think on nature is to theorise; and difficult it is not to be led on by the continuities of natural phenomena to theories which appear forced and unintelligible to those who have not pursued the same path of thought; which, moreover, if allowed to gain an undue influence, seduce us from that truth which is the sole object of our pursuit.

Where to draw the line—where to say thus far we may go, and no farther, in any particular class of analogies or relations which Nature presents to us; how far to follow the progressive indications of thought, and where to resist its allurements—is a question of degree which must depend upon the judgement of each individual or of each class of thinkers; yet it is consolatory that thought is seldom expended in vain.

I have throughout endeavoured to discard the hypotheses of subtle or occult entities; if in this endeavour some of my views have been adopted upon insufficient data, I still hope that this essay will not prove valueless.

The conviction that the so-called imponderables are modes of motion, will, at all events, lead the observer of natural phenomena to look for changes in these affections, wherever the intimate structure of matter is changed; and, conversely, to seek for changes in matter, either temporary or permanent, whenever it is affected by these forces. I believe he will seldom do this in vain. It was not until I had long reflected on the subject, that I ventured to publish my views; their publication may induce others to think on their subject-matter. They are not put forward with the same objects, nor do they aim at the same elaboration of detail, as memoirs on newly-discovered physical facts: they purport to be a method of mentally regarding known facts, some few of which I have myself made known on other occasions, but the great mass of which have been accumulated by the labours of others, and are admitted as established truths. Every one has a right to view these facts through any medium he thinks fit to employ, but some theory must exist in the minds of those who reflect upon the many new phenomena which have recently, and more particularly during the present century, been discovered. It is by a generalised or connected view of past acquisitions in natural knowledge that deductions can best be drawn as to the probable character of the results to be anticipated. It is a great assistance in such investigations to be intimately convinced that no physical phenomenon can stand alone: each is inevitably connected with anterior changes, and as inevitably productive of consequential changes, each with the other, and all with time and space; and, either in tracing back these antecedents or following up their consequents, many new phenomena will be discovered, and many existing phenomena hitherto believed distinct will be connected and explained: explanation is, indeed, only relation to something more familiar, not more known—i.e. known as to causative or creative agencies. In all phenomena the more closely they are investigated the more are we convinced that, humanly speaking, neither matter nor force can be created or annihilated, and that an essential cause is unattainable.—Causation is the will, Creation the act, of God.

ADDRESS

OF

WILLIAM ROBERT GROVE, ESQ.

PRESIDENT OF THE BRITISH ASSOCIATION FOR THE ADVANCEMENT
OF SCIENCE, NOTTINGHAM, 1866.

If our rude predecessors, who at one time inhabited the caverns which surround this town, could rise from their graves and see it in its present state, it may be doubtful whether they would have sufficient knowledge to be surprised.

The machinery, almost resembling organic beings in delicacy of structure, by which are fabricated products of world-wide reputation, the powers of matter applied to give motion to that machinery, are so far removed from what must have been the conceptions of the semi-barbarians to whom I have alluded, that they could not look on them with intelligent wonder.

Yet this immense progress has all been effected step by step, now and then a little more rapidly than at other times; but, viewing the whole course of improvement, it has been gradual, though moving in an accelerated ratio. But it is not merely in those branches of natural knowledge which tend to improvements in economical arts and manufactures, that science has made great progress. In the study of our own planet and the organic beings with which it is crowded, and in so much of the universe as vision, aided by the telescope, has brought within the area of observation, the present century has surpassed any antecedent period of equal duration.

It would be difficult to trace out all the causes which have led to the increase of observational and experimental knowledge.

Among the more thinking portion of mankind the gratification felt by the discovery of new truths, the expansion of faculties, and extension of the boundaries of knowledge, have been doubtless a sufficient inducement to the study of nature; while, to more practical minds, the reality, the certainty, and the progressive character of the acquisitions of natural science, and the enormously increased means which its applications give, have impressed its importance as a minister to daily wants and a contributor to ever-increasing material comforts, luxury, and power.

Though by no means the only one, yet an important cause of the rapid advance of science is the growth of associations for promoting the progress either of physical knowledge generally, or of special branches of it. Since the foundation of the Royal Society, now more than two centuries ago, a vast number of kindred societies have sprung up in this country and in Europe. The advantages conferred by these societies are manifold; they enable those who are devoted to scientific research, to combine, compare, and check their observations, to assist, by the thoughts of several minds, the promotion of the inquiry undertaken; they contribute from a joint purse to such efforts as their members deem most worthy; they afford a means of submitting to a competent tribunal notices and memoirs, and of obtaining for their authors and others, by means of the discussions which ensue, information given by those best informed on the particular subject; they enable the author to judge whether it is worth his while to pursue the subjects he has brought forward, and they defray the expense of printing and publishing such researches as are thought meritorious.

These advantages, and others might be named, pertain to the Association the thirty-sixth meeting of which we are this evening assembled to inaugurate; but it has, from its intermittent and peripatetic character, advantages which belong to none of the societies which are fixed as to their locality.

Among these are the novelty and freshness of an

annual meeting, which, while it brings together old members of the Association, many of whom only meet on this occasion, always adds a quota of new members, infusing new blood, and varying the social character of our meetings.

The visits of distinguished foreigners, whom we have previously known by reputation, is one of the most delightful and improving of the results. The wide field of inquiry, and the character of communications made to the Association, including all branches of natural knowledge, and varying from simple notices of an interesting observation or experiment, to the most intricate and refined branches of scientific research, is another valuable characteristic.

Lastly, perhaps, the greatest advantage resulting from the annual visits of this great parliament to new localities is that, while it imparts fresh local knowledge to the visitors, it leaves behind stimulating memories, which rouse into permanent activity dormant or timid minds—an effect which, so far from ceasing with the visit of the Association, frequently begins when that visit terminates.

Every votary of physical science must be anxious to see it recognised by those institutions of his country which can to the greatest degree promote its cultivation and reap from it the greatest benefit. You will probably agree with me that the principal educational establishments on the one hand, and on the other the Government, in many of its departments, are the institutions which may best fulfil these conditions. The more early the mind is trained to a pursuit of any kind, the deeper and more permanent are the impressions received, and the more service can be rendered by the students.

> ' Quo semel est imbuta recens, servabit odorem
> Testa diu.'

Little can be achieved in scientific research without an acquaintance with it in youth; you will rarely find an instance of a man who has attained any eminence in science who has not commenced its study at a very early period of life. Nothing, again, can tend more to the promotion of science than the exertions of those who have early acquired the ἦθος resulting from a scientific education. I desire to make no complaint of the tardiness with which science has been received at our public schools, and, with some exceptions, at our universities. These great establishments have their roots in historical periods, and long time and patient endeavour is requisite before a new branch of thought can be grafted with success on a stem to which it is exotic. Nor should I ever wish to see the study of languages, of history, of all those refined associations which the past has transmitted to us, neglected; but there is room for both. It is sad to see the number of so-called educated men who, travelling by railway, voyaging by steamboat, consulting the almanac for the time of sunrise or full-moon, have not the most elementary knowledge of a steam-engine, a barometer, or a quadrant; and who will listen with a half-confessed faith to the most idle predictions as to weather or cometic influences, while they are in a state of crass ignorance as to the cause of the trade winds, or the form of a comet's path. May we hope that the slight infiltration of scientific studies, now happily commenced, will extend till it occupies its fair space in the education of the young, and that those who may be able learnedly to discourse on the Eolic digamma will not be ashamed of knowing the principles on which the action of an air-pump, an electrical machine, or a telescope, depends, and will not, as Bacon complained of his contemporaries, despise such knowledge as something mean and mechanical.

To assert that the great departments of Government should encourage physical science may appear a truism, and yet it is but of late that it has been seriously done ; now, the habit of consulting men of science on important questions of national interest is becoming a recognised practice, and in a time, which may seem long to individuals, but is short in the history of a nation, a more definite sphere of usefulness for national purposes will, I have no doubt, be provided for those duly qualified men who may be content to give up the more tempting study of abstract science for that of its practical applications. In this respect the report of the Kew Committee for this year affords a subject of congratulation to those whom I have the honour to address. The Kew Observatory, the petted child of the British Association, may possibly become an important national establishment; and if so, while it will not, I trust, lose its character of a home for untrammelled physical research, it will have superadded some of the functions of the Meteorological Department of the Board of Trade, with a staff of skilful and experienced observers.

This is one of the results which the general growth of science, and the labours of this Association in particular, have produced; but I do not propose on this occasion to recapitulate the special objects attained by the Association; this has been amply done by several of my predecessors; nor shall I confine my address to the progress made in physical science since the time when my most able and esteemed friend and predecessor addressed you at Birmingham. In the various reports and communications which will be read at your sections, details of every step which has been made in science since our last meeting will be brought to your notice, and I have no doubt fully and freely discussed.

I purpose, with your kind permission, to submit to you certain views of what has within a comparatively recent period been accomplished by science, what have been the steps leading to the attained results, and what, as far as we may fairly form an opinion, is the general character pervading modern discovery.

It seems to me that the object we have in view would be more nearly approached, by each president, chosen as they are in succession as representing different branches of science, giving on these occasions either an account of the progress of the particular branch of science he has cultivated, when that is not of a very limited and special character, or enouncing his own view of the general progress of science; and though this will necessarily involve much that belongs to recent years, the confining a president to a mere *résumé* of what has taken place since our last meeting would, I venture with diffidence to think, limit his means of usefulness, and render his discourse rather an annual register than an instructive essay.

I need not dwell on the commonplace but yet important topics of the material advantages resulting from the application of science; I will address myself to what, in my humble judgement, are the lessons we have learned, and the probable prospects of improved natural knowledge.

One word will give you the key to what I am about to discourse on; that word is *continuity*, no new word, and used in no new sense, but perhaps applied more generally than it has hitherto been. We shall see, unless I am much mistaken, that the developement of observational, experimental, and even deductive knowledge is either attained by steps so extremely small as to form really a continuous ascent; or, when distinct results apparently separate from any co-ordinate phenomena have been attained, that then, by the subsequent progress of science, intermediate links have been discovered uniting the apparently segregated instances with other more familiar phenomena. We shall see that the more we investigate, the more we find that in existing phenomena graduation from the like to the seemingly unlike prevails, and in the changes which take place in time, gradual progress is, and apparently must be, the course of nature.

Let me now endeavour to apply this view to the recent progress of some of the more prominent branches of science.

In Astronomy, from the time when the earth was considered a flat plain bounded by a flat ocean —when the sun, moon, and stars were regarded as lanterns to illuminate this plain—each successive discovery has brought with it similitudes and analogies between this earth and many of the objects of the universe, with which our senses, aided by instruments, have made us acquainted. I pass, of course, over those discoveries which have established the Copernican system as applied to our sun, its attendant planets, and their satellites. The proofs, however, that gravitation is not confined to our solar system, but pervades the universe, have received many confirmations by the labours of members of this Association; I may name those who have held the office of President, Lord Rosse, Lord Wrottesley, and Sir J. Herschel, the two latter having devoted special attention to the orbits of double stars, the former to those probably more recent systems called nebulæ. Double stars seem to be orbs analogous to our own sun and revolving round their common centre of gravity in a conic-section curve, as do the planets with which we are more intimately acquainted; but the nebulæ present more difficulty, and some doubt has been expressed whether gravitation, such as we consider it, acts with those bodies (at least those exhibiting a spiral form) as it does with us; possibly some other modifying influence may exist, our present ignorance of which gives rise to the apparent difficulty. There is, however, another class of observations quite recent in its importance, and which has formed a special subject of contribution to the reports and transactions of this Association; I allude to those on Meteorites, at which our lamented member, and to many of us our valued friend, Prof. Baden Powell, assiduously laboured, for investigations into which a committee of this Association is formed, and a series of star-charts for enabling observers of shooting-stars to record their observations was laid before the last meeting of the Association by Mr. Glaisher.

It would occupy too much of your time to detail the efforts of Bessel, Schwinke, the late Sir J. Lubbock, and others, as applied to the formation of star-charts for aiding the observation of meteorites which Mr. Alexander Herschel, Mr. Brayley, Mr. Sorby, and others are now studying.

Dr. Olmsted explained the appearance of a point from which the lines of flight of meteors seem to radiate, as being the perspective vanishing point of their parallel or nearly parallel courses appearing to an observer on the earth as they approach it. The uniformity of position of these radiant points, the many corroborative observations on the direction, the distances, and the velocities of these bodies, the circumstance that their paths intersect the earth's orbit at certain definite periods, and the total failure of all other theories which have been advanced, while there is no substantial objection to this, afford evidence almost amounting to proof that these are cosmical bodies moving in the interplanetary space by gravitation round the sun, and some perhaps round planets. This view gives us a new element of continuity. The universe would thus appear not to have the extent of empty space formerly attributed to it, but to be studded between the larger and more visible masses with smaller planets, if the term be permitted to be applied to meteorites.

Observations are now made at the periods at which meteors appear in greatest numbers—at Greenwich by Mr. Glaisher, at Cambridge by Prof. Adams, and at Hawkhurst by Mr. Alexander Herschel—and every preparation is made to secure as much accuracy as can, in the present state of knowledge, be secured for such observations.

The number of known asteroids, or bodies of a smaller size than what are termed the ancient planets, has been so increased by numerous discoveries, that instead of seven we now count eighty-eight as the number of recognised planets —a field of discovery with which the name of Hind will be ever associated.

The smallest of these is only twenty or thirty miles in diameter, indeed cannot be accurately measured, and if we were to apply the same scrutiny to other parts of the heavens as has been applied to the zone between Mars and Jupiter, it is no far-fetched speculation to suppose that, in addition to asteroids and meteorites, many other bodies exist until the space occupied by our solar system becomes filled up with planetary bodies varying in size from that of Jupiter (1240 times larger in volume than the earth) to that of a cannon-ball or even a pistol-bullet.

The researches of Leverrier on the intra-mercurial planets come in aid of these views; and another half-century may, and not improbably will, enable us to ascertain that the now seemingly vacant interplanetary spaces are occupied by smaller bodies which have hitherto escaped observation, just as the asteroids had until the time of Olbers and Piazzi. But the evidence of continuity as pervading the universe does not stop at telescopic observation; chemistry and physical optics bring us new proofs. Those meteoric bodies which have from time to time come so far within reach of the earth's attraction as to fall upon its surface, give on analysis metals and oxides similar to those which belong to the structure of the earth —they come as travellers bringing specimens of minerals from extra-terrestrial regions.

In a series of papers recently communicated to the French Academy, M. Daubrée has discussed the chemical and mineralogical character of meteorites as compared with the rocks of the earth. He finds that the similarity of terrestrial rocks to meteorites increases as we penetrate deeper into the earth's crust, and that some of the deep-seated minerals have a composition and characteristics almost identical with meteorites [olivine, herzolite, and serpentine, for instance, closely resemble them]; that as we approach the surface, rocks having similar components with meteorites are found, but in a state of oxidation, which necessarily much modifies their mineral character, and which, by involving secondary oxygenised compounds, must also change their chemical constitution. By experiments he has succeeded in forming from terrestrial rocks substances very much resembling meteorites. Thus close relationship, though by no means identity, is established between this earth and those wanderers from remote regions, some evidence, though at present incomplete, of a common origin.

Surprise has often been expressed that, while the mean specific gravity of this globe is from five to six times that of water, the mean specific gravity of its crust is barely half as great. It has long seemed to me that there is no ground for wonder here. The exterior of our planet is to a considerable depth oxidated; the interior is in all probability free from oxygen, and whatever bodies exist there are in a reduced or deoxidated state; if so, their specific gravity must necessarily be higher than that of their oxides or chlorides, &c.; we find, moreover, that some of the deep-seated minerals have a higher specific gravity than the average of those on the surface; olivine, for instance, has a specific gravity of 3·3. There is therefore no à priori improbability that the mean specific gravity of the earth should notably exceed that of its surface; and if we go further, and suppose the interior of the earth to be formed of the same ingredients as the exterior, minus oxygen, chlorine, bromine, &c., a specific gravity of 5 to 6 would not be an unlikely one. Many of the elementary bodies entering largely into the formation of the earth's crust are as light or lighter than water—for instance, potassium, sodium, &c.; others, such as sulphur, silicon, aluminium, have from two to three times its specific gravity; others, again, as iron, copper, zinc, tin, seven to nine times; while others, lead, gold, platinum, &c., are much more dense—but, speaking generally, the more dense are the least numerous. There seems no improbability in a mixture of such substances producing a mean specific gravity of from 5 to 6, although it by no means follows, indeed the probability is rather the other way, that the proportions of the substances in the interior of the earth are the same as on the exterior. It might be worth the labour to ascertain the mean specific gravity of all the known minerals on the earth's surface, averaging them in the ratios in which, as far as our knowledge goes, they quantitatively exist, and assuming them to exist without the oxygen, chlorine, &c., with which they are, with some rare exceptions, invariably combined on the surface of the earth: great assistance to the knowledge of the probable constitution of the earth might be derived from such an investigation.

While chemistry, analytic and synthetic, thus

aids us in ascertaining the relationship of our planet to meteorites, its relation in composition to other planets, to the sun, and to more distant suns and systems, is aided by another science, viz. optics.

That light passing from one transparent medium to another should carry with it evidence of the source from which it emanates, would, until lately, have seemed an extravagant supposition; but probably (could we read it) everything contains in itself a large portion of its own history.

I need not detail to you the discoveries of Kirchhoff, Bunsen, Miller, Huggins, and others; they have been dilated on by my predecessor. Assuming that spectrum analysis is a reliable indication of the presence of given substances by the position of transverse bright lines exhibited when they are burnt, and of transverse dark lines when light is transmitted through their vapours, though Plücker has shown that with some substances these lines vary with temperature, the point of importance in the view I am presenting to you is, that while what may be called comparatively neighbouring cosmical bodies exhibit lines identical with many of those shown by the components of this planet, as we proceed to the more distant appearances of the nebulæ we get but one or two of such lines, and we get one or two new bands not yet identified with any known to be produced by substances on this globe.

Within the last year Mr. Huggins has added to his former researches observations on the spectrum of a comet (comet 1 of 1866), the nucleus of which shows but one bright line, while the spectrum formed by the light of the coma is continuous, seeming to show that the nucleus is gaseous, while the coma would consist of matter in a state of minute division shining by reflected light: whether this be solid, liquid, or gaseous is doubtful, but the author thinks it is in a condition analogous to that of fog or cloud. The position in the spectrum of the bright line furnished by the nucleus is the same as that of nitrogen, which line is also shown in some of the nebulæ.

But the most remarkable achievement by spectrum analysis is the record of observations on a temporary star which has shone forth this year in the constellation of the northern crown about a degree S.E. of the star ϵ. When it was first seen, May 12, it was nearly equal in brilliancy to a star of the second magnitude; when observed by Mr. Huggins and Dr. Miller, May 16, it was reduced to the third or fourth magnitude. Examined by these observers with the spectroscope, it gave a spectrum which they state was unlike that of any celestial body they had examined.

The light was compound and had emanated from two different sources. One spectrum was analogous to that of the sun, viz. formed by the light of an incandescent solid or liquid photosphere which had suffered absortion by the vapours of an envelope cooler than itself. The second spectrum consisted of a few bright lines, which indicated that the light by which it was formed was emitted by matter in the state of luminous gas. The observers consider that, from the position of two of the bright lines, the gas must be probably hydrogen, and from their brilliancy compared with the light of the photosphere the gas must have been at a very high temperature. They imagine the phenomena to result from the burning of hydrogen with some other element, and that from the resulting temperature the photosphere is heated to incandescence.

There is strong reason to believe that this star is one previously seen by Argelander and Sir J. Herschel, and that it is a variable star of long or irregular period; it is also notable that some of its spectrum lines correspond with those of several variable stars. The time of its appearance was too short for any attempt to ascertain its parallax; it would have been important if it could even have been established that it is not a near neighbour, as the magnitude of such a phenomenon must depend upon its distance. I forbear to add any speculations as to the cause of this most singular phenomenon; however imperfect the knowledge given us by these observations, it is a great triumph to have caught this fleeting object, and obtained permanent records for the use of future observers.

It would seem as if the phenomenon of gradual change obtained towards the remotest objects with which we are at present acquainted, and that the further we penetrate into space the more unlike to those we are acquainted with become the objects of our examination—sun, planets, meteorites, earth similarly though not identically constituted, stars differing from each other and from our system, and nebulæ more remote in space and differing more in their characters and constitution.

While we can thus to some extent investigate the physical constitution of the most remote visible substances, may we not hope that some further insight as to the constitution of the nearest, viz. our own satellite, may be given us by this class of researches? The question whether the moon possesses any atmosphere may still be regarded as unsolved. If there be any, it must be exceedingly small in quantity and highly attenuated. Calculations, made from occultations of stars, on the apparent differences of the semidiameter of the bright and dark moon give an amount of difference which might indicate a minute atmosphere, but which Mr. Airy attributes to irradiation.

Supposing the moon to be constituted of similar materials to the earth, it must be, to say the least, doubtful whether there is oxygen enough to oxidate the metals of which she is composed; and if not, the surface which we see must be metallic, or nearly so. The appearance of her craters is not unlike that seen on the surface of some metals, such as bismuth, or, according to Professor Phillips, silver, when cooling from fusion and just previous to solidifying; and it might be a fair subject of enquiry whether, if there be any coating of oxide on the surface, it may not be so thin as not to disguise the form of the congealed metallic masses, as they may have set in cooling from igneous fusion. M. Chacornac's recent observations lead him to suppose that many of the lunar craters were the result of a single explosion, which raised the surface as a bubble and deposited its débris around the orifice of eruption.

The eruptions on the surface of the moon clearly did not take place at one period only, for at many parts of the disk craters may be seen encroaching on and disfiguring more ancient craters, sometimes to the extent of three or four successive displacements: two important questions might, it seems to me, be solved by an attentive examination of such portions of the moon. By observing carefully with the most powerful telescopes the character of the ridges thus successively formed, the successive states of the lunar surface at different epochs might be elucidated; and secondly, as on the earth we should look for actual volcanic action at those points where recent eruptions have taken place, so on the moon the more recently active points being ascertained by the successive displacement of anterior formations, it is these points which should be examined for existing disruptive disturbances. Metius and Fabricius might be cited as points of this character, having been found by M. Chacornac to present successive displacements and to be perforated by numerous channels or cavities. M. Chacornac considers that the seas, as they are called, or smoother portions of the lunar surface have at some time made inroads on anteriorly formed craters; if so, a large portion of the surface of the moon must have been in a fused, liquid, semiliquid, or alluvial state long after the solidifying of other portions of it. It would be difficult to suppose that this state was one of igneous fusion, for this could hardly exist over a large part of the surface without melting up the remaining parts; on the other hand, the total absence of any signs of water, and of any, or, if any, only the most attenuated, atmosphere, would make it equally difficult to account for a large diluvial formation.

Some substances, like mercury on this planet, might have remained liquid after others had solidified; but the problem is one which needs more examination and study before any positive opinion can be pronounced.

I cannot pass from the subject of lunar physics

without recording the obligation we are under to our late President for his most valuable observations and for his exertion in organising a band of observers devoted to the examination of this our nearest celestial neighbour, and to Mr. Nasmyth and Mr. De la Rue for their important graphical and photographical contributions to this subject. The granular character of the sun's surface observed by Mr. Nasmyth in 1860 is also a discovery which ought not to be passed over in silence.

Before quitting the subject of astronomy I cannot avoid expressing a feeling of disappointment that the achromatic telescope, which has rendered such notable service to this science, still retains in practice the great defect which was known a century ago at the time of Hall and Dollond, namely, the inaccuracy of definition arising from what was termed the irrationality of the spectrum, or the incommensurate divisions of the spectra formed by flint and crown glass.

The beautiful results obtained by Blair have remained inoperative from the circumstance that evaporable liquids being employed between the lenses, a want of permanent uniformity in the instrument was experienced; and notwithstanding the high degree of perfection to which the grinding and polishing object-glasses has been brought by Clarke, Cooke, and Mertz, notwithstanding the greatly improved instrumental manufacture, the defect to which I have adverted remains unremedied and an eyesore to the observer with the refracting telescope.

We have now a large variety of different kinds of glass formed from different metallic oxides. A list of many such was given by M. Jacquelain a few years back; the last specimen which I have seen is a heavy high refracting glass formed from the metal thallium by M. Lamy. Among all these, could not two or three be selected which, having appropriate refracting and dispersing powers, would have the coloured spaces of their respective spectra if not absolutely in the same proportions, at all events much more nearly so than those of flint and crown glass? Could not, again, oily or resinous substances having much action on the more refrangible rays of the spectrum, such as castor oil, canada balsam, &c., be made use of in combination with glass lenses to reduce if not annihilate this signal defect? This is not a problem to the solution of which there seems any insuperable difficulty; the reason why it has not been solved is, I incline to think, that the great practical opticians have no time at their disposal to devote to long tentative experiments and calculations, and on the other hand the theoretic opticians have not the machinery and the skill in manipulation requisite to give the appropriate degree of excellence to the materials with which they experiment; yet the result is worth labouring for, as, could the defect be remedied, the refracting telescope would make nearly as great an advance upon its present state as the achromatic did on the single lens refractor.

While gravitation, physical constitution, and chemical analysis by the spectrum show us that matter has similar characteristics in other worlds than our own, when we pass to the consideration of those other attributes of matter which were at one time supposed to be peculiar kinds of matter itself, or, as they were called, imponderables, but which are now generally, if not universally, recognised as forces or modes of motion, we find the evidence of continuity still stronger.

When all that was known of magnetism was that a piece of steel rubbed against a particular mineral had the power of attracting iron, and, if freely suspended, of arranging itself nearly in a line with the earth's meridian, it seemed an exceptional phenomenon. When it was observed that amber, if rubbed, had the temporary power of attracting light bodies, this also seemed something peculiar and anomalous. What are now magnetism and electricity? forces so universal, so apparently connected with matter as to become two of its invariable attributes, and that to speak of matter not being capable of being affected by these forces would seem almost as extravagant as to speak of matter not being affected by gravitation.

So with light, heat, and chemical affinity, not merely is every form of matter with which we are acquainted capable of manifesting all these modes of force, but so-called matter supposed incapable of such manifestations would to most minds cease to be matter.

Further than this it seems to me (though, as I have taken an active part for many years, now dating from a quarter of a century, in promoting this view, I may not be considered an impartial judge) that it is now proved that all these forces are so invariably connected *inter se* and with motion as to be regarded as modifications of each other, and as resolving themselves objectively into motion, and subjectively into that something which produces or resists motion, and which we call force.

I may perhaps be permitted to recal a forgotten experiment, which nearly a quarter of a century ago I showed at the London Institution, an experiment simple enough in itself, but which then seemed to me important from the consequences to be deduced from it, and the importance of which will be much better appreciated now than then.

A train of multiplying wheels ended with a small metallic wheel which, when the train was put in motion, revolved with extreme rapidity against the periphery of the next wheel, a wooden one. In the metallic wheel was placed a small piece of phosphorus, and as long as the wheels revolved the phosphorus remained unchanged, but the moment the last wheel was stopped, by moving a small lever attached to it, the phosphorus burst into flame. My object was to show that while motion of the mass continued, heat was not generated, but that when this was arrested, the force continuing to operate, the motion of the mass became heat in the particles. The experiment differed from that of Rumford's cannon-boring and Davy's friction of ice in showing that there was no heat while the motion was unresisted, but that the heat was dependent on the motion being impeded or arrested. We have now become so accustomed to this view, that whenever we find motion resisted we look to heat, electricity, or some other force as the necessary and inevitable result.

It would be out of place here, and treating of matters too familiar to the bulk of my audience, to trace how by the labours of Oersted, Seebeck, Faraday, Talbot, Daguerre, and others, materials have been provided for the generalisation now known as the correlation of forces or conservation of energy, while Davy, Rumford, Seguin, Mayer, Joule, Helmholtz, Thomson, and others (among whom I would not name myself, were it not that I may be misunderstood and supposed to have abandoned all claim to a share in the initiation of this, as I believe, important generalisation) have carried on the work; and how, sometimes by independent and, as is commonly the case, nearly simultaneous deductions, sometimes by progressive and accumulated discoveries, the doctrine of the reciprocal interaction, of the quantitative relation, and of the necessary dependence of all the forces has, I think I may venture to say, been established.

If magnetism be, as it is proved to be, connected with the other forces or affections of matter, if electrical currents always produce, as they are proved to do, lines of magnetic force at right angles to their lines of action, magnetism must be cosmical, for where there is heat and light there is electricity, and consequently magnetism. Magnetism, then, must be cosmical and not merely terrestrial. Could we trace magnetism in other planets and suns as a force manifested in axial or meridional lines, *i. e.* in lines cutting at right angles the curves formed by their rotation round an axis, it would be a great step; but it is one hitherto unaccomplished. The apparent coincidences between the maxima and minima of solar spots, and the decennial or undecennial periods of terrestrial magnetic intensity, though only empirical at present, might tend to lead us to a knowledge of the connection we are seeking; and the President of the Royal Society considers that an additional epoch of coincidence has arrived, making the fourth decennial period; but some doubt is thrown upon these coincidences by the magnetic observations made at Greenwich Observatory. In a paper published in the 'Transactions of the Royal Society,' 1863, the Astronomer Royal says, speaking of results extending over seventeen years, there is no appearance of decennial cycle in the recurrence of great magnetic disturbances; and Mr. Glaisher last year, in the physical section of this Association, stated that after persevering examination he had been unable to trace any connection between the magnetism of the earth and the spots on the sun.

Mr. Airy, however, in a more recent paper, suggests that currents of magnetic force having reference to the solar hour are detected, and seem to produce vortices or circular disturbances, and he invites further cooperative observation on the subject, one of the highest interest, but at present remaining in great obscurity.

One of the most startling suggestions as to the consequence resulting from the dynamical theory of heat is that made by Mayer, that by the loss of *vis viva* occasioned by friction of the tidal waves, as well as by their forming, as it were, a drag upon the earth's rotatory movement, the velocity of the earth's rotation must be gradually diminishing, and that thus, unless some undiscovered compensatory action exist, this rotation must ultimately cease, and changes hardly calculable take place in the solar system.

M. Delaunay considers part of the acceleration of the moon's mean motion which is not at present accounted for by planetary disturbances, to be due to the gradual retardation of the earth's rotation; to which view, after an elaborate investigation, the Astronomer Royal has given his assent.

Another most interesting speculation of Mayer is that with which you are familiar, viz. that the heat of the sun is occasioned by friction or percussion of meteorites falling upon it: there are some difficulties, not perhaps insuperable, in this theory. Supposing such cosmical bodies to exist in sufficient numbers they would, as they revolve round the sun, fall into it, not as an aërolite falls upon the earth directly by an intersection of orbits, but by the gradual reduction in size of the orbits, occasioned by a resisting medium; some portion of force would be lost, and heat generated in space by friction against such medium; when they arrive at the sun they would, assuming them, like the planets, to have revolved in the same direction, all impinge in a definite direction, and we might expect to see some symptoms of such in the sun's photosphere; but though this is in a constant state of motion, and the direction of these movements has been carefully investigated by Mr. Carrington and others, no such general direction is detected; and M. Faye, who some time ago wrote a paper pointing out many objections to the theory of solar heat being produced by the fall of meteoric bodies into the sun, has recently investigated the proper motions of sun spots, and believes he has removed certain apparent anomalies and reduced their motions to a certain regularity in the motion of the photosphere, attributable to some general action arising from the internal mass of the sun.

It might be expected that comets, bodies so light and so easily deflected from their course, would show some symptoms of being acted on by gravitation, were such a number of bodies to exist in or near their paths as are presupposed in the mechanical theory of solar heat.

Assuming the undulatory theory of light to be true, and that the motion which constitutes light is transmitted across the interplanetary spaces by a highly elastic ether, then unless this motion is confined to one direction, unless there be no interference, unless there be no viscosity, as it is now termed, in the medium, and, consequently, no friction, light must lose something in its progress from distant luminous bodies, that is to say, must lose something as light; for, as all reflecting minds are now convinced that force cannot be annihilated, the force is not lost, but its mode of action is changed. If light, then, is lost as light (and the observations of Struvé seem to show this to be so, that, in fact, a star may be so far distant that it can never be seen in consequence of its luminous emissions becoming extinct), what becomes of the transmitted force lost as light, but existing in some other form? So with heat: our sun, our earth, and planets are constantly radiating heat into space, so in all probability are the other suns, the stars, and their attendant planets. What becomes of the heat thus radiated into space? If the universe have no limit, and it is difficult to conceive one, heat and light should be everywhere uniform; and yet more is given off than is received by each cosmical body, for otherwise night would be as light and as warm as day. What becomes of the enormous force thus apparently non-recurrent in the same form? Does it return as palpable motion? Does it move or contribute to move suns and planets? And can it be conceived as a force similar to that which Newton speculated on as universally repulsive and capable of being substituted for universal attraction? We are in no position at present to answer such questions as these; but I know of no problem in celestial dynamics more deeply interesting than this, and we may be no further removed from its solution than the predecessors of Newton were from the simple dynamical relation of matter to matter which that potent intellect detected and demonstrated.

Passing from extra terrestrial theories to the narrower field of molecular physics, we find the doctrine of correlation of forces steadily making its way. In the Bakerian Lecture for 1863, Mr. Sorby shows, not perhaps a direct correlation of mechanical and chemical forces, but that when, either by solution or by chemical action, a change in volume of the resulting substance as compared with that of its separate constituents is effected, the action of pressure retards or promotes the change, according as the substance formed would occupy a larger or smaller space than that occupied by its separate constituents; the application of these experiments to geological inquiries as to subterranean changes which may have taken place under great pressure is obvious, and we may expect to form compounds under artificial compression which cannot be found under normal pressure.

In a practical point of view the power of converting one mode of force into another is of the highest importance, and with reference to a subject which at present, somewhat prematurely, perhaps, occupies men's minds, viz. the prospective exhaustion of our coal-fields, there is every encouragement derivable from the knowledge that we can at will produce heat by the expenditure of other forces; but, more than that, we may probably be enabled to absorb or store up, as it were, diffused energy—for instance, Berthelot has found that the potential energy of formate of potash is much greater than that of its proximate constituents, caustic potash and carbonic oxide. This change may take place spontaneously and at ordinary temperatures, and by such change carbonic oxide becomes, so to speak, reinvested with the amount of potential energy which its carbon possessed before uniting with oxygen, or, in other words, the carbonic oxide is raised as a force-possessor to the place of carbon by the direct absorption or conversion of heat from surrounding matter.

Here we have as to force-absorption, an analogous result to that of the formation of coal from carbonic acid and water; and though this is a mere illustration, and may never become economical on a large scale, still it and similar examples may calm apprehension as to future means of supplying heat, should our present fuel become exhausted. As the sun's force, spent in times long past, is now returned to us from the coal which was formed by that light and heat, so the sun's rays, which are daily wasted, as far as we are concerned, on the sandy deserts of Africa, may hereafter, by chemical or mechanical means, be made to light and warm the habitations of the denizens of colder regions. The tidal wave is, again, a large reservoir of force hitherto almost unused.

The valuable researches of Professor Tyndall on radiant heat, afford many instances of the power of localising, if the term be permitted, heat which would otherwise be dissipated.

The discoveries of Graham, by which atmospheric air, drawn through films of caoutchouc, leaves behind half its nitrogen, or, in other words, becomes richer by half in oxygen, and hence has a much increased potential energy, not only show a most remarkable instance of physical molecular action, merging into chemical, but afford us indications of means of storing up force, much of the force used in working the aspirator being capable at any period, however remote, of being evolved by burning the oxygen with a combustible.

What changes may take place in our modes of applying force before the coal-fields are exhausted it is impossible to predict. Even guesses at the probable period of their exhaustion are uncertain. There is a tendency to substitute for smelting in metallurgic processes, liquid chemical action, which of course has the effect of saving fuel; and the waste of fuel in ordinary operations is enormous, and can be much economised by already known processes. It is true that we are, at present, far from seeing a practical mode of replacing that granary of force the coal-fields; but we may with confidence rely on invention being in this case, as in others, born of necessity, when the necessity arises.

I will not further pursue this subject; at a time

when science and civilisation cannot prevent large tracts of country being irrigated by human blood in order to gratify the ambition of a few restless men, it seems an over-refined sensibility to occupy ourselves with providing means for our descendants in the tenth generation to warm their dwellings or propel their locomotives.

Two very remarkable applications of the convertibility of force have been recently attained by the experiments of Mr. Wilde and Mr. Holz; the former finds that, by conveying electricity from the coils of a magneto-electric machine to an electromagnet, a considerable increase of electrical power may be attained, and by applying this as a magneto-electric machine to a second, and this in turn to a third electro-magnetic apparatus, the force is largely augmented. Of course, to produce this increase, more mechanical force must be used at each step to work the magneto-electric machines; but provided this be supplied, there hardly seems a limit to the extent to which mechanical may be converted into electrical force.

Mr. Holz has contrived a Franklinic electrical machine, in which a similar principle is manifested. A varnished glass plate is made to revolve in close proximity to another plate having two or more pieces of card attached, which are electrified by a bit of rubbed glass or ebonite; the moment this is effected, a resistance is felt by the operator who turns the handle of the machine, and the slight temporary electrization of the card converts into a continuous flood of intense electricity the force supplied by the arm of the operator.

These results offer great promise of extended application; they show that, by a mere formal disposition of matter, one force may be converted into another, and that not to the limited extent hitherto attained, but to an extent co-ordinate, or nearly so, with the increased initial force, so that, by a mere change in the arrangement of apparatus, a means of absorbing and again eliminating in a new form a given force may be obtained to an indefinite extent. As we may, in a not very distant future, need, for the daily uses of mankind, heat, light, and mechanical force, and find our present resources exhausted, the more we can invent new modes of conversion of forces, the more prospect we have of practically supplying such want. It is but a month from this time that the greatest triumph of force-conversion has been attained. The chemical action generated by a little salt-water on a few pieces of zinc will now enable us to converse with inhabitants of the opposite hemisphere of this planet, and

'Put a girdle round about the earth in forty minutes.'

The Atlantic Telegraph is an accomplished fact.

In physiology very considerable strides are being

made by studying the relation of organised bodies to external forces; and this branch of enquiry has been promoted by the labours of Carpenter, Bence Jones, Playfair, E. Smith, Frankland, and others. Vegetables acted on by light and heat, decompose water, ammonia, and carbonic acid, and transform them into, among other substances, oxalate of lime, lactic acid, starch, sugar, stearine, urea, and ultimately albumen; while the animal reverses the process, as does vegetable decay, and produces from albumen, urea, stearine, sugar, starch, lactic acid, oxalate of lime, and ultimately ammonia, water, and carbonic acid.

As, moreover, heat and light are absorbed, or converted in forming the synthetic processes going on in the vegetable, so conversely heat and sometimes light is given off by the living animal; but it must not be forgotten that the line of demarcation between a vegetable and an animal is difficult to draw, that there are no single attributes which are peculiar to either, and that it is only by a number of characteristics that either can be defined.

The series of processes above given may be simulated by the chemist in his laboratory; and the amount of labour which a man has undergone in the course of twenty-four hours may be approximately arrived at by an examination of the chemical changes which have taken place in his body; changed forms in matter indicating the anterior exercise of dynamical force. That muscular action is produced or supported by chemical change would probably now be a generally accepted doctrine; but while many have thought that muscular power is derived from the oxidation of albuminous or nitrogenised substances, several recent researches seem to show that the latter is rather an accompaniment than a cause of the former, and that it is by the oxidation of carbon and hydrogen compounds that muscular force is supplied. Traube has been prominent in advancing this view, and experiments detailed in a paper published this year by two Swiss professors, Drs. Fick and Wislicenus, which were made by and upon themselves in an ascent of the Faulhorn, have gone far to confirm it. Having fed themselves, before and during the ascent, upon starch, fat, and sugar, avoiding all nitrogenised compounds, they found that the consumption of such food was amply sufficient to supply the force necessary for their expedition, and that they felt no exhaustion. By appropriate chemical examination they ascertained that there was no notable increase in the oxidation of the nitrogenised constituents of the body. After calculating the mechanical equivalents of the combustion effected, they state, as their first conclusion, that 'the burning of protein substances cannot be the only source of muscular power, for we have here two cases in which men performed more measurable work than the equivalent of the amount of heat which, taken at a most absurdly high figure, could be calculated to result from the burning of the albumen.'

They further go on to state that, so far from the oxidation of albuminous substances being the only source of muscular power, 'the substances by the burning of which force is generated in the muscles are not the albuminous constituents of those tissues, but non-nitrogenous substances, either fats or hydrates of carbon,' and that the burning of albumen is not in any way concerned in the production of muscular power.

We must not confuse the question of the food which forms and repairs muscle and gives permanent capability of muscular force with that which supplies the requisites for temporary activity; no doubt the carnivora are the most powerfully constituted animals, but the chamois, gazelle, &c., have great temporary capacity for muscular exertion, though their food is vegetable; for concentrated and sustained energy, however, they do not equal the carnivora; and with the domestic graminivora we certainly find that they are capable of performing more continuous work when supplied with those vegetables which contain the greatest quantity of nitrogen.

These and many similar classes of research show that in chemical enquiries, as in other branches of science, we are gradually relieving ourselves of hypothetical existences, which certainly had the advantage that they might be varied to suit the requirements of the theorist.

Phlogiston, as Lavoisier said with a sneer, was sometimes heavy, sometimes light; sometimes fire in a free state, sometimes combined; sometimes passing through glass vessels, sometimes retained by them; which by its protean changes explained causticity and non-causticity, transparency and opacity, colours and their absence. As phlogiston and similar creations of the mind have passed away, so with hypothetic fluids, imponderable matters, specific ethers, and other inventions of entities made to vary according to the requirements of the theorist, I believe the day is approaching when these will be dispensed with, and when the two fundamental conceptions of matter and motion will be found sufficient to explain physical phenomena.

The facts made known to us by geological enquiries, while on the one hand they afford striking evidence of continuity, on the other, by the breaks in the record, may be used as arguments against it. The great question once was, whether these chasms represent sudden changes in the formation of the earth's crust, or whether they arise from dislocations occasioned since the original deposition of strata or from gradual shifting of the areas of submergence. Few geologists of the present day would, I imagine, not adopt the latter alternatives. Then comes a second question, whether, when the geological formation is of a continuous character, the different characters of the fossils represent absolutely permanent varieties, or may be explained by gradual modifying changes.

Professor Ansted, summing up the evidence on this head as applied to one division of stratified rocks, writes as follows:—' Palæontologists have endeavoured to separate the Lias into a number of subdivisions, by the Ammonites, groups of species of those shells being characteristic of different zones. The evidence on this point rests on the assumption of specific differences being indicated by permanent modifications of the structure of the shell. But it is quite possible that these may mean nothing more than would be due to some change in the conditions of existence. Except between the Marlstone and the Upper Lias there is really no palæontological break, in the proper sense of the words; alterations of form and size consequent on the occurrence of circumstances more or less favourable, migration of species, and other well-known causes, sufficiently account for many of those modifications of the form of the shell that have been taken as specific marks. This view is strengthened by the fact that other shells and other organisms generally show no proof of a break of any importance except at the point already alluded to.'

But, irrespectively of another deficiency in the geological record, which will be noticed presently, the physical breaks in the stratification make it next to impossible to fairly trace the order of succession of organisms by the evidence afforded by their fossil remains. Thus there are nine great breaks in the Palæozoic series, four in the Secondary, and one in the Tertiary, besides those between Palæozoic and Secondary and Secondary and Tertiary respectively. Thus in England there are sixteen important breaks in the succession of strata, together with a number of less important interruptions. But although these breaks exist, we find pervading the works of many geologists a belief, resulting from the evidence presented to their minds, sometimes avowed, sometimes unconsciously implied, that the succession of species bears some definite relation to the succession of strata. Thus Professor Ramsay says, that 'In cases of superposition of fossiliferous strata, in proportion as the species are more or less continuous, that is to say, as the break in the succession of life is partial or complete, so was the time that elapsed between the close of the lower and the commencement of the upper strata a shorter or a longer interval. The break in life may be indicated not only by a difference in species, but yet more importantly by the absence of older and appearance of newer allied or unallied genera.'

Indications of the connection between cosmical studies and geological researches are dawning on us: there is, for instance, some reason to believe that we can trace many geological phenomena to our varying rotation round the sun; thus more than thirty years ago Sir J. Herschel proposed an explanation of the changes of climate on the earth's surface as evidenced by geological phenomena, founded on the changes of eccentricity in the earth's orbit.

He said he had entered on the subject 'impressed with the magnificence of that view of geological revolutions which regards them rather as regular and necessary efforts of great and general causes, than as resulting from a series of convulsions and catastrophes regulated by no laws and reducible to no fixed principles.'

As the mean distance of the earth from the sun is nearly invariable, it would seem at first sight that the mean annual supply of light and heat received by the earth would also be invariable; but according to his calculations it is inversely proportional to the minor axis of the orbit: this would give less heat when the eccentricity of the earth's orbit is approaching towards or at its minimum. Mr. Croll has recently shown reason to believe that the climate, at all events in the circumpolar and temperate zones of the earth, would depend on whether the winter of a given region occurred when the earth at its period of greatest eccentricity was in aphelion or perihelion —if the former, the annual average of temperature would be lower, if the latter, it would be higher than when the eccentricity of the earth's orbit were less or approached more nearly to a circle. He calculates the difference in the amount of heat at the period of maximum eccentricity of the earth's orbit to be as 19 to 26, according as the winter would take place when the earth was in aphelion or in perihelion. His reason may be briefly stated thus: assuming the mean annual heat to be the same, whatever the eccentricity of orbit, yet if the extremes of heat and cold in winter and summer be greater, a colder climate will prevail, for there will be more snow and ice accumulated in the cold winter than the hot summer can melt, a result aided by the shelter from the sun's rays produced by the vapour suspended in consequence of the aqueous evaporation; hence we should get glacial periods, when the orbit of the earth is at its greatest eccentricity, at those parts of the earth's surface where it is winter when the earth is in aphelion; carboniferous or hot periods where it is winter in perihelion; and normal or temperate periods when the eccentricity of orbit is at a minimum; all these would gradually slide into each other, and would produce at long distant periods alternations of cold and heat, several of which we actually observe in geological records.

If this theory be borne out, we should approximate to a test of the time which has elapsed between different geological epochs. Mr. Croll's computation of this would make it certainly not less than 100,000 years since the last glacial epoch, a time not very long in geological chronology—probably it is much more.

When we compare with the old theories of the earth, by which the apparent changes on its surface were accounted for by convulsions and cataclysms, the modern view inaugurated by Lyell, your former President, and now, if not wholly, at all events to a great extent adopted, it seems strange that the referring past changes to similar causes to those which are now in operation should have remained uninvestigated until the present century; but with this, as with other branches of knowledge, the most simple is frequently the latest view which occurs to the mind. It is much more easy to invent a *Deux ex machinâ* than to trace out the influence of slow continuous change; the love of the marvellous is so much more attractive than the patient investigation of truth, that we find it to have prevailed almost universally in the early stages of science.

In astronomy we had crystal spheres, cycles,

and epicycles; in chemistry the philosopher's stone, the elixir vitæ, the archæus or stomach demon, and phlogiston; in electricity the notion that amber possessed a soul, and that a mysterious fluid could knock down a steeple. In geology a deluge or a volcano was supplied. In palæontology a new race was created whenever theory required it: how such new races began, the theorist did not stop to enquire.

A curious speculator might say to a palæontologist of even recent date, in the words of Lucretius,

> 'Nam neque de cœlo cecidisse animalia possunt
> Nec terrestria de salsis exisse lacunis.
> * * * * *
> ·E nihilo si crescere possent,
> (Tum) fierent juvenes subito ex infantibus parvis,
> E terrâque exorta repente arbusta salirent;
> Quorum nil fieri manifestum est, omnia quando
> Paulatim crescunt, ut par est, semine certo,
> Crescentesque genus servant'

—which may be thus freely paraphrased: 'You have abandoned the belief in one primæval creation at one point of time, you cannot assert that an elephant existed when the first saurians roamed over earth and water. Without, then, in any way limiting Almighty power, if an elephant were created without progenitors, the first elephant must, in some way or other, have physically arrived on this earth. Whence did he come? did he fall from the sky (*i.e.* from interplanetary space)? did he rise moulded out of a mass of amorphous earth or rock? did he appear out of the cleft of a tree? If he had no antecedent progenitors, some such beginning must be assigned to him.' I know of no scientific writer who has, since the discoveries of geology have become familiar, ventured to present in intelligible terms any definite notion of how such an event could have occurred: those who do not adopt some view of continuity are content to say God willed it; but would it not be more reverent and more philosophical to enquire by observation and experiment, and to reason from induction and analogy, as to the probabilities of such frequent miraculous interventions?

I know I am touching on delicate ground, and that a long time may elapse before that calm enquiry after truth which it is the object of associations like this to promote can be fully attained; but I trust that the members of this body are sufficiently free from prejudice, whatever their opinions may be, to admit an enquiry into the general question whether what we term species are and have been rigidly limited, and have at numerous periods been created complete and unchangeable, or whether, in some mode or other, they have not gradually and indefinitely varied, and whether the changes due to the influence of surrounding circumstances, to efforts to accommodate themselves to surrounding changes, to what is called natural selection, or to the necessity of yielding to superior force in the struggle for existence, as maintained by our illustrious countryman Darwin, and other causes, have not so modified organisms as to enable them to exist under changed conditions. I am not going to put forward any theory of my own, I am not going to argue in support of any special theory, but having endeavoured to show how, as science advances, the continuity of natural phenomena becomes more apparent, it would be cowardice not to present some of the main arguments for and against continuity as applied to the history of organic beings.

As we detect no such phenomenon as the creation or spontaneous generation of vegetables and animals which are large enough for the eye to see without instrumental assistance, as we have long ceased to expect to find a Plesiosaurus spontaneously generated in our fish-pond, or a Pterodactyle in our pheasant-cover, the field of this class of research has become identified with the field of the microscope, and at each new phase the investigation has passed from a larger to a smaller class of organisms. The question whether among the smallest and apparently the most elementary forms of organic life the phenomenon of spontaneous generation obtains, has recently formed the subject of careful experiment and animated discussion in France. If it could be found that organisms of a complex character were generated without progenitors out of amorphous matter, it might reasonably be argued that a similar mode of creation might obtain in regard to larger organisms. Although we see no such phenomenon as the formation of an animal such as an elephant, or a tree such as an oak, excepting from a parent which resembles it, yet if the microscope revealed to us organisms, smaller but equally complex, so formed without having been reproduced, it would render it not improbable that such might have been the case with larger organic beings. The controversy between M. Pasteur and M. Pouchet has led to a very close investigation of this subject, and the general opinion is that when such precautions are taken as exclude from the substance submitted to experiment all possibility of germs from the atmosphere being introduced, as by passing the air which is to support the life of the animalculæ through tubes heated to redness and other precautions, no formation of organisms takes place. Some experiments of Mr. Child's, communicated to the Royal Society during the last year, again throw doubt on the negative results obtained by M. Pasteur; so that the question may be not finally determined, but the balance of experiment and opinion is against spontaneous generation.

One argument presented by M. Pasteur is well worthy of remark, viz. that in proportion as our means of scrutiny become more searching, heterogeny, or the developement of organisms without generation from parents of similar organism, has been gradually driven from higher to lower forms of life, so that if some apparent exceptions still exist they are of the lowest and simplest forms, and these exceptions may probably be removed, as M. Pasteur considers he has removed them, by a more searching investigation.

If it be otherwise, if heterogeny obtains at all, few will not now admit that at present the result of the most careful experiments shows it to be confined to the more simple organic structures, and that all the progressive and more highly developed forms are, as far as the most enlarged experience shows, generated by reproduction.

The great difficulty which is met with at the threshold of enquiry into the origin of species is the definition of species; in fact, species can hardly be defined without begging the question in dispute.

Thus if species be said to be a perseverance of type incapable of blending itself with other types, or, which comes nearly to the same thing, incapable of producing by union with other types offspring of an intermediate character which can again reproduce, we arrive at this result, that whenever the advocate of continuity shows a blending of what had been hitherto deemed separate species, the answer is, they were considered separate species by mistake, they do not now come under the definition of species, because they interbreed.

The line of demarcation is thus *ex hypothesi* removed a step further, so that unless the advocate of continuity can, on his side, prove the whole question in dispute, by showing that all can directly or by intermediate varieties reproduce, he is defeated by the definition itself of species.

On the other hand, if this, or something in fact amounting to it, be not the definition of species— if it be admitted that distinct species can, under certain favourable conditions, produce intermediate offspring capable of reproduction, then continuity in some mode or other is admitted.

The question then takes this form. Are there species or are there not? Is the word to be used as signifying a real, natural distinction, or as a mere convenient designation applied to subdivisions having a permanence which will probably outlive man's discussions on the subject, but not an absolute fixity? The same question, in a wider sense, and taking into consideration a much longer time, would be applicable to genera and families.

Actual experiment has done little to elucidate the question, nor, unless we can suppose the experiments continued through countless generations, is it likely to contribute much to its solution. We must therefore have recourse to the enlarged experience or induction from the facts of geology, palæontology, and physiology, aided by analogy from the laws of action which nature evidences in other departments.

The doctrine of gradual succession is hardly yet formularised, and though there are some high authorities for certain modifications of such view, the preponderance of authority would necessarily be on the other side. Geology and palæontology are recent sciences, and we cannot tell what the older authors would have thought or written had the more recently discovered facts been presented to their view. Authority, therefore, does not much help us on this question.

Geological discoveries seemed, in the early period of the science, to show complete extinction of certain species and the appearance of new ones, great gaps existing between the characteristics of the extinct and the new species. As science advanced, these were more or less filled up, and the difficulty in the first instance of admitting unlimited modification of species would seem to have arisen from the comparison of the extreme ends of the scale where the intermediate links or some of them were wanting.

To suppose a Zoophyte the progenitor of a Mammal, or to suppose at some particular period of time a highly developed animal to have come out of nothing, or suddenly grown out of inorganic matter, would appear at first sight equally extravagant hypotheses. As an effort of Almighty creative power, neither of these alternatives presents more difficulty than the other; but as we have no means of ascertaining how creative power worked, but by an examination and study of the works themselves, we are not likely to get either view proved to ocular demonstration. A single phase in the progress of natural transmutation would probably require a term far transcending all that embraced by historical records; and on the other hand, it might be said, sudden creations, though taking place frequently, if viewed with reference to the immensity of time involved in geological periods, may be so rare with reference to our experience, and so difficult of clear authentication, that the non-observation of such instances cannot be regarded as absolute disproof of their possible occurrence.

The more the gaps between species are filled up by the discovery of intermediate varieties the stronger becomes the argument for transmutation and the weaker that for successive creations, because the former view then becomes more and more consistent with experience, the latter more discordant from it. As undoubted cases of variation, more or less permanent, from given characteristics, are produced by the effects of climate, food, domestication, &c., the more species are increased by intercalation, the more the distinctions slide down towards those which are within the limits of such observed deviations; while on the other hand, to suppose the more and more frequent recurrence of fresh creations out of amorphous matter is a multiplication of miracles or special interventions not in accordance with what we see of the uniform and gradual progress of nature, either in the organic or inorganic world. If we were entitled to conclude that the progress of discovery would continue in the same course, and that species would become indefinitely multiplied, the distinctions would become infinitely minute, and all lines of demarcation would cease, the polygon would become a circle, the succession of points a line. Certain it is that the more we observe the more we increase the subdivision of species, and consequently the number of these supposed creations; so that new creations become innumerable, and yet of these we have no one well-authenticated instance, and in no other observed operation of nature have we seen this want of continuity, these frequent *per saltum* deviations from uniformity, each of which is a miracle.

The difficulty of producing intermediate offspring from what are termed distinct species and the infecundity in many instances of hybrids are used as strong arguments against continuity of succession;

on the other hand, it may be said long-continued variation through countless generations has given rise to such differences of physical character that reproduction is difficult in some cases, and in others impossible.

Suppose, for instance, M to represent a parent-race whose offspring by successive changes through eons of time have divaricated, and produced on the one hand a species A, and on the other a species Z, the changes here have been so great that we should never expect directly to reproduce an intermediate between A and Z. A and B on the one hand, and Y and Z on the other, might reproduce; but to regain the original type M, we must not only retrocede through all the intermediates, but must have similar circumstances recalled in an inverse order at each phase of retrogression, conditions which it is obviously impossible to fulfil. But though among the higher forms of organic structure we cannot retrace the effects of time and reproduce intermediate types, yet among some of the lower forms we find it difficult to assign any line of specific demarcation; thus as a result of the very elaborate and careful investigations of Dr. Carpenter on Foraminifera, he states, 'It has been shown that a very wide range of variation exists among Orbitolites, not merely as regards external form, but also as to plan of developement; and not merely as to the shape and aspect of the entire organism, but also with respect to the size and configuration of its component parts. It would have been easy, by selecting only the most divergent types from amongst the whole series of specimens which I have examined, to prefer an apparently substantial claim on behalf of these to be accounted as so many distinct species. But after having classified the specimens which could be arranged around these types, a large proportion would yet have remained, either presenting characters intermediate between those of two or more of them, or actually combining those characters in different parts of their fabric; thus showing that no lines of demarcation can be drawn across any part of the series that shall definitely separate it into any number of groups, each characterised by features entirely peculiar to itself.'

At the conclusion of his enquiry he states—

I. The range of variation is so great among Foraminifera as to include not merely the differential characters which systematists proceeding upon the ordinary methods have accounted specific, but also those upon which the greater part of the genera of this group have been founded, and even in some instances those of its orders.

II. The ordinary notion of species as assemblages of individuals marked out from each other by definite characters that have been genetically transmitted from original proto-types similarly distinguished, is quite inapplicable to this group; since even if the limits of such assemblages were extended so as to include what elsewhere would be accounted genera, they would still be found so intimately connected by gradational links that definite lines could not be drawn between them.

III. The only natural classification of the vast aggregate of diversified forms which this group contains will be one which ranges them according to their direction and degree of divergence from a small number of principal family types; and any subordinate grouping of genera and species which may be adopted for the convenience of description and nomenclature must be regarded merely as assemblages of forms characterised by the nature and degree of the modifications of the original type, which they may have respectively acquired in the course of genetic descent from a common ancestry.

IV. Even in regard to these family types it may fairly be questioned whether analogical evidence does not rather favour the idea of their derivation from a common original than that of their primitive distinctness.

Mr. H. Bates, when investigating 'The Lepidoptera of the Amazon Valley,' may almost be said to have witnessed the origin of some species of Butterflies, so close have been his observations on the habits of these animals that have led to their variation and segregation, so closely do the results follow his observations, and so great is the difficulty of otherwise accounting for any of the observed facts.

In the numerous localities of the Amazon region certain gregarious species of Butterfly (*Heliconidea*) swarm in incredible numbers, almost outnumbering all the other butterflies in the neighbourhood; the species in the different localities being different, though often to be distinguished by a very slight shade.

In these swarms are to be found, in small numbers, other species of butterflies belonging to as many as ten different genera, and even some moths; and these intruders, though they structurally differ *in toto* from the swarms they mingle with, and from one another, mimic the *Heliconideæ* so closely in colours, habits, mode of flight, &c., that it is almost impossible to distinguish the intruders from those they mingle with. The obvious benefit of this mimicry is safety, the intruders hence escaping detection by predatory animals.

Mr. Bates has extended his observations to the habits of life, food, variations and geographical range of the species concerned in these mimetic phenomena, and finds in every case corroborative evidence of every variety and species being derivative, the species being modified from place to place to suit the peculiar form of *Heliconidea* stationed there.

Mr. Wallace has done similar service to the derivative theory by his observations and writings on the Butterflies and Birds of the Malay Archipelago, adducing instances of mimetic resemblances strictly analogous to the above; and adding in further illustration a beautiful series of instances where the form of the wing of the same butterfly is so modified in various islets as to produce changes in their mode of flight that tend to the conservation of the variety by aiding its escape when chased by birds or predacious insects.

He has also adduced a multitude of examples of geographical and representative species, races, and varieties, forming so graduated a series as to render it obvious that they have had a common origin.

The effect of food in the formation and segregation of races and of certain groups of insects has been admirably demonstrated by Mr. B. D. Walsh, of North America.

Mr. McDonnell has been led to the discovery of a new organ in electric fishes from the application of the theory of descent, and Dr. Fritz Müller has published numerous observations showing that organs of very different structure may, through the operation of natural selection, acquire very similar and even identical functions. Sir John Lubbock's diving hymenopterous insect affords a remarkable illustration of analogous phenomena; it dives by the aid of its wings, and is the only insect of the vast order it belongs to that is at all aquatic.

The discovery of the Eozoon is of the highest importance in reference to the derivative hypothesis, occurring as it does in strata that were formed at a period inconceivably antecedent to the presupposed introduction of life upon the globe, and displacing the argument derived from the supposition that at the dawn of life a multitude of beings of high organisation were simultaneously developed (in the Silurian and Cambrian strata).

Professor A. De Candolle, one of the most distinguished continental botanists, has, to some extent, abandoned the tenets held in his 'Géographie Botanique,' and favours the derivative hypothesis in his paper on the variation of oaks; following up a paper, by Dr. Hooker, on the oaks of Palestine, showing that some sixteen of them are derivative, he avows his belief that two-thirds of the 300 species of this genus, which he himself describes, are provisional only.

Dr. Hooker, who had only partially accepted the derivative hypothesis propounded before the publication of 'The Origin of Species through Natural Selection,' at the same time declining the doctrine of special creation, has since then cordially adopted the former, and illustrated its principles by applying them to the solution of various botanical questions: first, in reference to the flora of Australia, the anomalies of which he appears to explain satisfactorily by the application of these principles; and, latterly, in reference to the Arctic flora.

In the case of the Arctic flora, he believes that originally Scandinavian types were spread over the high northern latitudes; that these were driven southwards during the glacial period, when many of them changed their forms in the struggle that ensued with the displaced temperate plants; that on the returning warmth, the Scandinavian plants, whether changed or not, were driven again northwards and up to the mountains of the temperate latitudes, followed in both cases by series of pre-existing plants of the temperate Alps. The result is the present mixed Arctic flora, consisting of a basis of more or less changed and unchanged Scandinavian plants, associated in each longitude with representatives of the mountain flora of the more temperate regions to the south of them.

The publication of a previously totally unknown flora, that of the Alps of tropical Africa, by Dr. Hooker, has afforded a multitude of facts that have been applied in confirmation of the derivative hypothesis. This flora is found to have relationships with those of temperate Europe and North Africa, of the Cape of Good Hope, and of the mountains of tropical Madagascar and Abyssinia, that can be accounted for on no other hypothesis but that there has been ancient climatal connexion and some coincident or subsequent slight changes of specific character.

The doctrine of Cuvier, every day more and more borne out by observation, that each organ bears a definite relation to the whole of the individual, seems to support the view of indefinite variation. If an animal seeks its food or safety by climbing trees, its claws will become more prehensile, the muscles which act upon those claws must become more developed, the body will become agile by the very exercise which is necessary to it, and each portion of the frame will mould itself to the wants of the animal by the effect on it of the habits of the animal.

Another series of facts which present an argument in favour of gradual succession are the phases of resemblance to inferior orders which the embryo passes through in its developement, and the relations shown in what is termed the metamorphosis of plants; facts difficult to account for on the theory of frequent separate creations, but almost inevitable on that of gradual succession. So also, the existence of rudimentary and effete organs, which must either be referred to a *lusus naturæ* or to some mode of continuous succession.

The doctrine of typical nuclei seems only a mode of evading the difficulty; experience does not give us the types of theory, and, after all, what are these types? It must be admitted there are none such in reality; how are we led to the theory of them? simply by a process of abstraction from classified existences. Having grouped from natural similitudes certain forms into a class, we select attributes common to each member of the class, and call the assemblage of such attributes a type of the class. This process gives us an abstract idea, and we then transfer this idea to the Creator, and make Him start with that which our own imperfect generalisation has derived. It seems to me that the doctrine of types is, in fact, a concession to the theory of continuity or indefinite variability; for the admission that large groups have common characters shows, necessarily, a blending of forms within the scope of the group, which supports the view of each member being derived from some other member of it: can it be asserted that the assigned limits of such groups have a definite line of demarcation?

The condition of the earth's surface, or at least of large portions of it, has for long periods remained substantially the same; this would involve a greater degree of fixity in the organisms which have existed during such periods of little change than in those which have come into being during periods of more rapid transition; for, though rejecting catastrophes as the general *modus agendi* of nature, I am far from saying that the march of physical changes has been always perfectly uniform.

There have been doubtless what may be termed secular seasons, and there have been local changes of varying degrees of extent and permanence; from such causes organised beings would be more concentrated in certain directions than in others, the fixity of character being in the ratio of the fixity of condition. This would throw natural forms into certain groups which would be more prominent than others, like the colours of the rainbow, which present certain predominant tints though they merge into each other by insensible gradations.

While the evidence seems daily becoming stronger in favour of a derivative hypothesis as applied to the succession of organic beings, we are far removed from anything like a sufficient number of facts to show that, at all events within the existing geological periods capable of being investigated, there has been any great progression from a simpler or more embryonic to a more complex type.

Professor Huxley, though inclined to the derivative hypothesis, shows, in the concluding portion of his address to the Geological Society, 1862, a great number of cases in which, though there is abundant evidence of variation, there is none of progression. There are, however, several groups of Vertebrata in which the endoskeleton of the older presents a less ossified condition than that of the younger genera. He cites the Devonian

Ganoids, the Mesozoic Lepidosteidæ, the Palæozoic Sharks, and the more ancient Crocodilia and Lacertilia, and particularly the Pycnodonts and Labyrinthodonts, as instances of this when compared with their more recent representatives.

The records of life on the globe may have been destroyed by the fusion of the rocks, which would otherwise have preserved them, or by crystallisation after hydrothermal action. The earlier forms may have existed at a period when this planet was in course of formation, or being segregated or detached from other worlds or systems. We have not evidence enough to speculate on the subject, but by time and patience we may acquire it.

Were all the forms which have existed embalmed in rock the question would be solved; but what a small proportion of extinct forms is so preserved, and must be, if we consider the circumstances necessary to fossilise organic remains. On the dry land, unwashed by rivers and seas, when an animal or plant dies, it undergoes chemical decomposition which changes its form; it is consumed by insects, its skeleton is oxidised and crumbles into dust. Of the myriads of animals and vegetables which annually perish we find hardly an instance of a relic so preserved as to be likely to become a permanent fossil. So again in the deeper parts of the ocean, or of the larger lakes, the few fish there are perish and their remains sink to the bottom, and are there frequently consumed by other marine or lacustrine organisms or chemically decomposed. As a general rule, it is only when the remains are silted up by marine, fluviatile or lacustrine sediments that the remains are preserved. Geology, therefore, might be expected to keep for us mainly such organic remains as inhabited deltas or the margins of seas, lakes, or rivers; here and there an exception may occur, but the mass of preserved relics would be those of creatures so situated; and so we find it, the bulk of fossil remains consists of fish and amphibia, shell-fish form the major part of the geological museum, limestone and chalk rocks frequently consisting of little else than a congeries of fossil shells. Plants of reed or rush-like character, fish which are capable of inhabiting shallow waters, and saurian animals, form another large portion of geological remains.

Compare the shell-fish and amphibia of existing organisms with the other forms, and what a small proportion they supply; compare the shell-fish and amphibia of Palæontology with the other forms, and what an overwhelming majority they yield.

There is nothing, as Professor Huxley has remarked, like an extinct order of birds or mammals, only a few isolated instances. It may be said the ancient world possessed a larger proportion of fish and amphibia, and was more suited to their existence. I see no reason for believing this, at least to anything like the extent contended for; the fauna and flora now in course of being preserved for future ages would give the same idea to our successors.

Crowded as Europe is with cattle, birds, insects, &c., how few are geologically preserved! while the muddy or sandy margins of the ocean, the estuaries, and deltas are yearly accumulating numerous crustacea and mollusca, with some fishes and reptiles, for the study of future palæontologists.

If this position be right, then, notwithstanding the immense number of preserved fossils, there must have lived an immeasurably larger number of unpreserved organic beings, so that the chance of filling up the missing links, except in occasional instances, is very slight. Yet where circumstances have remained suitable for their preservation, many closely connected species are preserved—in other words, while the intermediate types in certain cases are lost, in others they exist. The opponents of continuity lay all stress on the lost and none on the existing links.

But there is another difficulty in the way of tracing a given organism to its parent forms, which, from our conventional mode of deducing genealogies, is never looked upon in its proper light.

Where are we to look for the remote ancestor of a given form? Each of us, supposing none of our progenitors to have intermarried with relatives, would have had at or about the period of the Norman Conquest upwards of a hundred million direct ancestors of that generation, and if we add the intermediate ancestors, double that number. As each individual has a male and female parent, we have only to multiply by two for each thirty years, the average duration of a generation, and it will give the above result.

Let any one assume that one of his ancestors at the time of the Norman Conquest was a Moor, another a Celt, and a third a Laplander, and that these three were preserved while all the others were lost, he would never recognise either of them as his ancestor, he would only have the one-hundred millionth of the blood of each of them, and as far as they were concerned there would be no perceptible sign of identity of race.

But the problem is more complex than that which I have stated; at the time of the Conquest there were hardly a hundred million people in Europe, it follows that a great number of the ancestors of the *propositus* must have intermarried with relations, and then the pedigree, going back to the time of the Conquest, instead of being represented by diverging lines, would form a network so tangled that no skill could unravel it; the law of probabilities would indicate that any two people in the same country, taken at hazard, would not have many generations to go back before they would find a common ancestor, who probably, could they have seen him or her in the life, had no traceable resemblance to either of them. Thus two animals of a very different form, and of what would be termed very different species, might have a common geological ancestor, and yet the skill of no comparative anatomist could trace the descent.

From the long-continued conventional habit of tracing pedigrees through the male ancestor, we forget in talking of progenitors that each individual has a mother as well as a father, and there is no reason to suppose that he has in him less of the blood of the one than of the other.

The recent discoveries in palæontology show us that Man existed on this planet at an epoch far anterior to that commonly assigned to him. The instruments connected with human remains, and indisputably the work of human hands, show that to these remote periods the term civilisation could hardly be applied—chipped flints of the rudest construction, probably, in the earlier cases, fabricated by holding an amorphous flint in the hand and chipping off portions of it by striking it against a larger stone or rock; then, as time suggested improvements, it would be more carefully shaped, and another stone used as a tool; then (at what interval we can hardly guess) it would be ground, then roughly polished, and so on—subsequently bronze weapons, and nearly the last before we come to historical periods, iron. Such an apparently simple invention as a wheel must, in all probability, have been far subsequent to the rude hunting-tools or weapons of war to which I have alluded.

A little step-by-step reasoning will convince the unprejudiced that what we call civilisation must have been a gradual process; can it be supposed that the inhabitants of Central America or of Egypt suddenly and what is called instinctively built their cities, carved and ornamented their monuments? If not, if they must have learned to construct such erections, did it not take time to acquire such learning, to invent tools as occasion required, contrivances to raise weights, rules or laws by which men acted in concert to effect the design? Did not all this require time? and if, as the evidence of historical times shows, invention marches with a geometrical progression, how slow must have been the earlier steps! If even now habit, and prejudice resulting therefrom, vested interests, &c., retard for some time the general application of a new invention, what must have been the degree of retardation among the comparatively uneducated beings which then existed?

I have of course been able to indicate only a few of the broad arguments on this most interesting subject; for detailed results the works of Darwin,

Hooker, Huxley, Carpenter, Lyell, and others must be examined. If I appear to lean to the view that the successive changes in organic beings do not take place by sudden leaps, it is, I believe, from no want of an impartial feeling; but if the facts are stronger in favour of one theory than another, it would be an affectation of impartiality to make the balance appear equipoised.

The prejudices of education and associations with the past are against this as against all new views; and while on the one hand a theory is not to be accepted because it is new and *primâ facie* plausible, still to this assembly I need not say that its running counter to existing opinions is not necessarily a reason for its rejection; the *onus probandi* should rest on those who advance a new view, but the degree of proof must differ with the nature of the subject. The fair question is, Does the newly-proposed view remove more difficulties, require fewer assumptions, and present more consistency with observed facts than that which it seeks to supersede? If so, the philosopher will adopt it, and the world will follow the philosopher—after many days.

It must be borne in mind that even if we are satisfied from a persevering and impartial enquiry that organic forms have varied indefinitely in time, the *causa causans* of these changes is not explained by our researches; if it be admitted that we find no evidence of amorphous matter suddenly changed into complex structure, still why matter should be endowed with the plasticity by which it slowly acquires modified structure is unexplained. If we assume that natural selection, or the struggle for existence, coupled with the tendency of like to reproduce like, gives rise to various organic changes, still our researches are at present uninstructive as to why like should produce like, why acquired characteristics in the parent should be reproduced in the offspring. Reproduction itself is still an enigma, and this great question may involve deeper thoughts than it would be suitable to enter upon now.

Perhaps the most convincing argument in favour of continuity which could be presented to a doubting mind would be the difficulty it would feel in representing to itself any *per saltum* act of nature. Who would not be astonished at beholding an oak tree spring up in a day, and not from seed or shoot? We are forced by experience, though often unconsciously, to believe in continuity as to all effects now taking place; if any one of them be anomalous we endeavour, by tracing its history and concomitant circumstances, to find its cause, *i.e.* to relate it to antecedent phenomena; are we then to reject similar enquiries as to the past? is it laudable to seek an explanation of present changes by observation, experiment, and analogy, and yet reprehensible to apply the same mode of investigation to the past history of the earth and of the organic remains embalmed in it?

If we disbelieve in sudden creations of matter or force, in the sudden formations of complex organisms now, if we now assign to the heat of the sun an action enabling vegetables to live by assimilating gases and amorphous earths into growing structures, why should such effects not have taken place in earlier periods of the world's history, when the sun shone as now, and when the same materials existed for his rays to fall upon?

If we are satisfied that continuity is a law of nature, the true expression of the action of Almighty Power, then, though we may humbly confess our inability to explain why matter is impressed with this tendency to gradual structural formation, we should cease to look for special interventions of creative power in changes which are difficult to understand, because, being removed from us in time, their concomitants are lost; we should endeavour from the relics to evoke their history, and when we find a gap not try to bridge it over with a miracle.

If it be true that continuity pervades all physical phenomena, the doctrine applied by Cuvier to the relations of the different parts of an animal to each other might be capable of great extension. All the phenomena of inorganic and organised matter might be expected to be so inter-related that the study of an isolated phenomenon would lead to a knowledge of numerous other phenomena with which it is connected. As the antiquary deduces from a monolith the tools, the arts, the habits, and epoch of those by whom it is wrought, so the student of science may deduce from a spark of electricity or ray of light the source whence it is generated; and by similar processes of reasoning other phenomena hitherto unknown may be deduced from their probable relation with the known. But, as with heat, light, magnetism, and electricity, though we may study the phenomena to which these names have been given, and their mutual relations, we know nothing of what they are; so, whether we adopt the view of natural selection, of effort, of plasticity, &c., we know not why organisms should have this *nisus formativus*, or why the acquired habit or exceptional quality of the individual should reappear in the offspring.

Philosophy ought to have no likes or dislikes, truth is her only aim; but if a glow of admiration be permitted to a physical enquirer, to my mind a far more exquisite sense of the beautiful is conveyed by the orderly developement, by the necessary inter-relation and inter-action of each element of the cosmos, and by the conviction that a bullet falling to the ground changes the dynamical conditions of the universe, than can be conveyed by mysteries, by convulsions, or by cataclysms.

The sense of understanding is to the educated more gratifying than the love of the marvellous, though the latter need never be wanting to the nature-seeker.

But the doctrine of continuity is not solely applicable to physical enquiries.

The same modes of thought which lead us to see continuity in the field of the microscope as in the universe, in infinity downwards as in infinity upwards, will lead us to see it in the history of our own race; the revolutionary ideas of the so-called natural rights of man, and *à priori* reasoning from what are termed first principles, are far more unsound and give us far less ground for improvement of the race than the study of the gradual progressive changes arising from changed circumstances, changed wants, changed habits. Our language, our social institutions, our laws, the constitution of which we are proud, are the growth of time, the product of slow adaptations, resulting from continuous struggles. Happily in this country practical experience has taught us to improve rather than to remodel; we follow the law of nature and avoid cataclysms.

The superiority of Man over other animals inhabiting this planet, of civilised over savage man, and of the more civilised over the less civilised, is proportioned to the extent which his thought can grasp of the past and of the future. His memory reaches further back, his capability of prediction reaches further forward in proportion as his knowledge increases. He has not only personal memory which brings to his mind at will the events of his individual life—he has history, the memory of the race; he has geology, the history of the planet; he has astronomy, the geology of other worlds. Whence does the conviction to which I have alluded, that each material form bears in itself the records of its past history, arise? Is it not from the belief in continuity? Does not the worn hollow on the rock record the action of the tide, its stratified layers the slow deposition by which it was formed, the organic remains imbedded in it the beings living at the times these layers were deposited, so that from a fragment of stone we can get the history of a period myriads of years ago? From a fragment of bronze we may get the history of our race at a period antecedent to tradition. As science advances our power of reading such history improves and is extended. Saturn's ring may help us to a knowledge of how our solar system developed itself, for it as surely contains that history as the rock contains the record of its own formation.

By this patient investigation how much have we already learned, which the most civilised of ancient human races ignored! While in ethics, in politics, in poetry, in sculpture, in painting, we have scarcely, if at all, advanced beyond the highest intellects of ancient Greece or Italy, how great are the steps we have made in physical science and its applications!

But how much more may we not expect to know?

We, this evening assembled, Ephemera as we are, have learned by transmitted labour, to weigh, as in a balance, other worlds larger and heavier than our own, to know the length of their days and years, to measure their enormous distance from us and from each other, to detect and accurately ascertain the influence they have on the movements of our world and on each other, and to discover the substances of which they are composed; may we not fairly hope that similar methods of research to those which have taught us so much may give our race further information, until problems relating not only to remote worlds, but possibly to organic and sentient beings which may inhabit them, problems which it might now seem wildly visionary to enunciate, may be solved by progressive improvements in the modes of applying observation and experiment, induction and deduction.

NOTES AND REFERENCES
TO THE CORRELATION OF PHYSICAL FORCES.

PAGE

7. THE reader who is curious as to the views of the ancients, regarding the objects of science, will find clues to them in the second book of ARISTOTLE's Physics, and in the first three books of the Metaphysics. See also the Timæus of PLATO, and RITTER's History of Ancient Philosophy, where a sketch of the Philosophy of LEUCIPPUS and DEMOCRITUS will be found.

9. BACON's Novum Organum, book ii. aph. 5 and 6.

11. HUME's Enquiry concerning Human Understanding, S. 7, London, 1768.

    BROWN's Enquiry into the Relations of Cause and Effect, London, 1835.

    The illustration I have used of a floodgate has been objected to, as being one to which the term cause would scarcely be applied, but after some consideration I have retained it; if cause be viewed only as sequence, it must be limited to sequence under given conditions or circumstances, and here, given the conditions, the sequence is invariable. I see no difference *quoad* the argument, between this illustration and that of BROWN of a lighted match and gunpowder (4th edit. p. 27), to which my reasoning would equally well apply.

    HERSCHEL's Discourse on the Study of Natural Philosophy, pp. 88 and 149.

13. Quarterly Review, vol. lxviii. p. 212.

    WHEWELL, On the Question 'Are Cause and Effect Successive or Simultaneous?' (Cambridge Philosophical Transactions, vol. vii. p. 319).

PAGE
14. HERSCHEL's Discourse, p. 93.
   AMPERE, Théorie des Phénomènes Electro-dynamiques, Memoirs in the Ann. de Chimie et de Physique, and works from 1820 to 1826, Paris.
21. LAMARCK, ' Sur la Matière du Son ' (Journal de Physique, vol. xlix. p. 397).
24. D'ALEMBERT, Traité de Dynamique, pp. 3 and 4, Paris, 1796.
28. BABBAGE, On the Permanent Impression of our Words and Actions on the Globe we inhabit, 9th Bridgewater Treatise, ch. ix.
32. MAYER, Annalen der Pharmacie Leibig und Wohler, May 1852.
   RANKINE, Measure of Moving Force. (The Engineer, October 26, 1866).
35. JOULE, On the Mechanical Equivalent of Heat (Phil. Trans. 1850, p. 61).
36. ERMAN, Influence of Friction upon Thermo-electricity (Reports of the British Association, 1845).
39. BECQUEREL, Dégagement de l'Electricité par Frottement, Traité de l'Electricité, tom. ii. p. 113 et seq.
40. SULLIVAN, Currents produced by the vibration of metals (Archiv. de l'Electricité, t. 10, p. 480). LEROUX, Vibrations arrested produce heat (Cosmos, March 30, 1860).
42. WHEATSTONE, On the Prismatic Decomposition of Electrical Light (Notices of Communications to the British Association, p. 11, 1835).
43. BACON, De Formâ Calidi, Nov. Org. book 2, aph. 20.
   RUMFORD, An Enquiry concerning the Source of Heat which is excited by Friction (Phil. Trans. p. 80, 1798).
   DAVY, On the Conversion of Ice into Water by Friction (West of England Contributions, p. 16).
   Of Heat or Calorific Repulsion (Elements of Chemical Philosophy, p. 69).
46. BADEN POWELL, On the Repulsive Power of Heat (Phil. Trans. 1834, p. 485).
   FRESNEL, Annales de Chimie, tom. xxix. pp. 57 and 107.

NOTES AND REFERENCES.   351

PAGE
47. MOSER, On Invisible Light (Taylor's Scientific Memoirs, vol. iii. pp. 461 and 465).
49. BLACK, On Latent Heat (Elements of Chemistry, p. 144 et passim, 1803).
51. The experiments of HENRY and DONNY have shown that the cohesion of liquids, as far as their antagonism to rupture goes, is much greater than has been generally believed. These experiments, however, make no difference in the view I have put forth, as, whatever be the character of the attraction, there is a molecular attraction to be overcome in changing bodies from the solid to the liquid state, which must require and exhaust force.

DONNY, Sur la Cohésion des Liquides (Mémoires de l'Académie Royale de Bruxelles, 1843).

HENRY, Proceedings of the American Philosophical Society, April 1844 (Silliman's Journal, vol. xlviii. p. 215).

56. THILORIER, Solidification de l'Acide carbonique (Ann. de Ch. et de Phys. tom. lx. p. 432).
59. I. WEDGWOOD, Thermometer for measuring the Higher Degrees of Heat (Phil. Trans. 1782, p. 305; and 1786, p. 390).

TYNDALL, On the Physical Properties of Ice (Phil. Trans. 1858, p. 211).

DESPRETZ, Recherches sur le Maximum de Densité de l'Eau pure et des Dissolutions aqueuses (Ann. de Ch. et de Ph. tom. lxx. p. 45, and tom. lxxiii. p. 296).

60. BIOT (Comptes rendus de l'Académie des Sciences, Paris 1850, p. 281). The experiments on circular polarisation by water were, I believe, by Dr. Leeson.
62. I. THOMPSON, Trans. R. S. Edin. vol. xvi. p. 575.

W. THOMPSON, Phil. Mag. August 1850, p. 123.

BUNSEN, Pogg. Ann. vol. lxxxi. p. 562; Ann. de Ch. et de Phys. vol. xxxv. p. 383. Effects of Pressure on the Freezing Point.

63. JOULE, Phil. Trans. 1852, p. 99.

Although, taking the phenomena as they are known

to exist, the mechanical laws may be deduced, yet in any physical conception of the nature of heat expansion by cold has always been a great stumbling-block to me, and I believe to many others.

63. DULONG and PETIT, and REGNAULT. See their Memoirs abstracted and referred to in Gmelin's Handbook of Chemistry, translated by Watts for the Cavendish Society, vol. i. p. 242 et seq.

65. WOOD, Phil. Mag. 1851, 1852.

67. SENARMONT, Conduction of Heat by Crystals (Gmelin's Handbook, vol. i. p. 222).

68. KNOBLAUCH, Ann. de Ch. et de Ph. vol. xxxvi. p. 124.
TYNDALL, Transmission of Heat through Organic Structures (Phil. Trans. vol. cxliii. p. 217).

71. GROVE, Electricity produced by approximating Metals; Report of a Lecture at the London Institution (Literary Gazette, 1843, p. 39).
GASSIOT, Phil. Mag. October 1844.
ROGET, On the Improbability of the Contact exciting Force: Treatise on Galvanism (Library of Useful Knowledge, S. 113).
FARADAY, Phil. Trans. 1840, p. 126.

73. MELLONI, Sur la Polarisation de la Chaleur: Recherches sur plusieurs Phénomènes calorifiques (Annales de Chimie et de Ph. tom. xlv. pp. 5–68; tom. xli. pp. 375–410; tom. xlviii. pp. 198, 218).
FORBES, On the Refraction and Polarisation of Heat (Transactions of the Royal Society of Edinburgh, vol. xiii. pp. 131, 168).

75. KIRCHOFF, Trans. Berlin Acad. 1861.
BALFOUR STEWART on the theory of Exchanges (Report British Association, 1861).

77. T. WEDGWOOD, On the Production of Light and Heat by different Bodies (Phil. Trans. vol. lxxxii. p. 272).

78. TYNDALL, Proc R. S. vol. iv. p. 487.

81. GROVE, On the Decomposition of Water into its Constituent Gases by Heat (Phil. Trans. 1847, p. 1).

NOTES AND REFERENCES. 353

PAGE

81. ROBINSON, On the Effect of Heat in lessening the Affinities of the Elements of Water (Transactions of the Royal Irish Academy, vol. xxi. p. 2).

83. GROVE, Water decomposed by Chlorine and Heat (Phil. Trans. 1847, p. 20).

87. CARNOT, Réflexions sur la Puissance motrice du Feu, Paris, 1824.

96. SEGUIN, Influence des Chemins de Fer, p. 378 et seq.

98. ROGERS, Consumption of Coal for Man power (Cosmos, vol. ii. p. 56).

102. MAYER and WATERSTON have suggested that solar heat may arise from the mechanical action of meteoric stones falling into the sun, and Mr. THOMPSON has written an elaborate paper on the subject (Trans. Brit. Assoc. 1853). If a number of gravitating bodies exist in the neighbourhood of the sun, and form, as is conjectured, the zodiacal light, it is difficult to conceive how comets as they approach this region steer clear of such bodies, and are not even deflected from their orbits.

For Mr. THOMPSON's various and valuable papers, see Phil. Mag. passim.

103. POISSON, Comptes rendus, Paris, January 30, 1837.

106. DUFAYE, SYMMER, WATSON, and FRANKLIN, Theories of Electric Fluid and Electric Fluids (Priestley's History of Electricity, pp. 429-441).

107. GROTTHUS, Sur la Décomposition de l'Eau et des Corps qu'elle tient en dissolution à aide de l'Electricité galvanique (Ann. de Chimie, tom. lviii. p. 54).

FARADAY, On the Question whether Electrolytes conduct without Decomposition (Proceedings of the Weekly Meetings of the Royal Institution, 1855).

GROVE (Comptes rendus, Paris, 1839).

108. FARADAY, On Induction as an Action of contiguous Particles (Phil. Trans. 1838, p. 30).

109. MATTEUCCI, Plates of Mica polarised by Electricity (De la Rive's Electricity, p. 140).

A A

PAGE

109. GROVE, Electrolysis across Glass (Phil. Mag. Aug. 1860).

110. KARSTEN on Electrical Figures (Archiv. de l'Elec. vols. ii. iii. and iv.)

111. GROVE, Etching Electrical Figures and transferring them to Collodion (Phil. Mag. January 1857).

113. FUSINIERI, Du Transport des Matières pondérables qui s'opère dans les Décharges électriques (Archives de l'Electricité; Supplément à la Bibliothèque universelle de Genève, tom. iii. p. 597).

114. GROVE, On the Voltaic Arc (Report of Lecture at the Royal Institution, Lit. Gaz. and Athenæum, Feb. 7, 1845; Phil. Trans. 1847, p. 16).

116 to 122. GROVE, On the Electro-chemical Polarity of Gases (Phil. Trans. 1852, p. 87).

121. FREMY and E. BECQUEREL, Oxygen changed to Ozone by the Electric Spark (Ann. de Ch. et de Phys. 1852). This subject and the nature of Ozone was first investigated by Dr. Schönbein. See also a paper by Mr. Brodie On the Conditions of certain Elements at the Moment of Chemical Change (Phil. Trans. 1850).

123, 124. Molecular Changes in Electrised Metals (NAIRNE, Phil. Trans. 1780, p. 334, and 1783, p. 223; GROVE, Electrical Mag. vol. i. p. 120; PELTIER, Archives de l'Electricité, vol. v. p. 182; FUSINIERI, id. p. 516).

125. WERTHEIM, Change in Elasticity of Metals by Electrisation (Ann. de Ch. et de Phys. vol. xii. p. 623; Arch. Elec. vol. iv. p. 490).

DUFOUR, Alteration in Tenacity of Metals by Electrisation (Bibl. univ. de Genève, Fev. 1855, p. 156).

126. MATTEUCCI, Conduction of Electricity by Crystals (Comptes rendus de l'Acad., Paris, March 5, 1855, p. 541).

128. E. BECQUEREL, Transmission of Electricity by heated Gases (Ann. de Ch. et de Phys. vol. xxxix. p. 355).

GROVE, Proceedings of the Royal Inst. (1854, p. 361).

BECQUEREL, Divergence of Gold-Leaves in Vacuo (Traité d'Electricité, vol. v.; part ii. p. 55).

NOTES AND REFERENCES.   355

PAGE
128. NEWTON, Thirty-first Query to the Optics.
130. GROVE, Particles of Metals and Metallic Oxides detached in Liquids by Electricity (Elec. Mag. vol. i. p. 119).
131. MATTEUCCI, Relations of Electricity and Nervous Force (Phil. Trans. 1845, p. 285, 1846, p. 497; Phénomènes physiques des Corps vivants, p. 305; Lezioni di Fisica, p. 360).
  GALVANI VOLTA MARIANINI et NOBILI on Physiological Effects of Electricity (Ann. de Ch. et de Phys. vols. 23, 25, 29, 38, 40, 43, 44, 56).
133. BECQUEREL, Chemical Changes by Friction (Traité de l'Elec. vol. v. part 1, p. 16).
139. DE LA RIVE, Heat of the Voltaic Pile (Bibl. univ. vol. xiii. p. 389).
  DAVY, On the Properties of Electrified Bodies in their relations to Conducting Powers and Temperature (Phil. Trans, 1821, p. 428).
140. GROVE, On the Effect of surrounding Media on Voltaic Ignition (Phil Trans. 1849, p. 49).
141. OERSTED, Expériences sur l'Effet du Conflit électrique sur l'Aiguille aimantée (Ann. de Ch. et de Phys. tom. xiv. p. 417).
142. COLERIDGE, Table Talk, vol. i. p. 65.
143. LENZ and JACOBI, Pogg. Ann. vol. xvlii. p. 403; Bulletin de l'Acad. St. Petersburg, 1839; Harris, Magnetism, part 2, p. 63.
  DAVY, Decomposition of the fixed Alkalies (Phil. Trans. 1808, p. 1).
  BECQUEREL Des Composés électro-chimiques (Traité de l'Electricité, vol. iii. c. 13).
  CROSSE, Transactions of the British Association, vol. v. p. 47; Proceedings of the Electrical Society, p. 320.
144. MALUS, Polarisation of Light by Reflexion (Mémoires d'Arcueil, tom. ii. p. 143).
  ARAGO, Circular Polarisation by Solids (Mémoires de l'Institut, 1811).

NOTES AND REFERENCES.

145. BIOT, Circular Polarisation by Liquids (Mémoires de l'Institut, 1817).

146. NIEPCE and DAGUERRE, Historique et Description des Procédés du Daguerréotype, Paris, 1839.
TALBOT, Photogenic Drawing and Calotype (Phil. Mag. March 1839, and August 1841).

149. HERSCHEL, Chemical Action of the Solar Spectrum on various Substances (Phil. Trans. 1840, p. 1, and 1842, p. 181).
HUNT, Researches on Light, London, 1844.

153. GROVE, Other Forces produced by Light (Lit. Gaz. January 1844).

154. GROVE, Influence of Light on the Polarised Electrode (Phil. Mag. December 1858).
SOMERVILLE (Mrs.), On the Magnetising Power of the more Refrangible Solar Rays (Phil. Trans. 1826, p. 132).
MORICHINI's experiments are given in Mrs. Somerville's paper.

155. HERSCHEL, On the Absorption of Light in Coloured Media viewed in connection with the Undulatory Theory (Phil. Mag. December 1833).
TYNDALL, Proc. R. I., vol. iv. p. 491.
SEEBECK, Heat of Coloured Rays (Brewster's Optics, p. 90).

156. KNOBLAUCH (Ann. de Ch. vol. xxxvi. p. 124, and Pogg. Ann. there referred to).

158. HERSCHEL, Epipolised Light (Phil. Trans. vol. cxxxv. pp. 143, 147).
STOKES, Change in Refrangibility of Light (Phil. Trans. vols. cxlii. cxliii.)

163. For the first enunciations of the Corpuscular and Undulatory Theories, see NEWTON's Optics, HOOKE's Micographia, and HUYGHENS' Tractatus de Lumine. See also BREWSTER's Optics, p. 138.

164. YOUNG, Lectures edited by Kelland, p. 359 et seq.; Phil. Trans. 1800, p. 126; HERSCHEL, Encyc. Metrop. art.

Light, pp. 450 and 738; NEWTON's Optics, p. 322; WHEWELL's Hist. Induc. Sc. vol. ii. p. 449; FOUCAULT, Comptes rendus, Paris, 1850, p. 65; HARRISON, Phil. Mag. November 1856; Camb. Phil. Trans.

167. SONDHAUSS, Refraction of Sound (Ann. de Ch. et de Phys. vol. xxxv. p. 505); DOVÉ, Polarisation of Sound (Cosmos, May 13, 1859).

176. PASTEUR, Rotation of Plane of Polarised Light by Solutions of Hemihedral Crystals (Ann. de Ch. et de Phys. vol. xxiv. p. 412).

179 to 183. WOLLASTON, Phil. Trans. 1822, p. 89; WHEWELL, Phil. of the Induct. Sc. vol. i. p. 419; WILSON, Trans. of the Roy. Soc. of Edin. vol. xvi. p. 79; Sir. W. HERSCHEL, Phil. Trans. 1793, p. 201, and 1801, p. 300; MORGAN, Phil. Trans. vol. lxxv. p. 272; DAVY, Phil. Trans. 1822, p. 64; Elements of Chemical Philosophy, p. 97; GASSIOT, Phil. Trans. 1859, p. 157.

184. Diminishing .Periods of Comets (Herschel's Outlines of Astronomy, p. 357).

187. Since writing the passage in the text, I find that STRUVE has been led, from his astronomical researches, to the conclusion that some light is lost in the interplanetary spaces. He gives as an approximation one per cent. as lost by the passage of light from a star of the first magnitude, assuming a mean or average distance (Etudes d'Astronomie Stellaire, 1847).

NEWTON, Thirtieth Query to the Optics.

190. FARADAY, Evolution of Electricity from Magnetism (Phil. Trans. 1832, p. 125).

193. FARADAY, Magnetic Condition of all Matter (Phil. Trans. 1846, p. 21; Phil. Mag. 1846, p. 249).

BECQUEREL, Ann. de Ch. et de Ph. tom. xxxvi. p. 337; Comptes rendus, Paris, 1846, p. 147; and 1850, p. 201.

194. FARADAY, On the Magnetisation of Light (Phil. Trans. 1846, p. 1).

PAGE
195. WARTMANN, Rotation of the Plane of Polarisation of Heat by Magnetism (Journal de l'Institut, No. 644).
PROVOSTAYE and DESSAINES, Ann. de Ch. et de Phys. October 1849.

196. HUNT, Influence of Magnetism on Molecular Arrangement (Phil. Mag. 1846, vol. xxviii. p. 1; Memoirs of the Geological Society, vol. i. p. 433).
WARTMANN, Phil. Mag. 1847, vol. xxx. p. 263.

197. GROVE, Experiment on Molecular Motion of a Magnetic Substance (Electrical Mag. 1845, vol. i. p. 601).

198. On the direct Production of Heat by Magnetism (Proceedings of the Royal Society, 1849, p. 826).

After this paper was communicated and ordered to be printed in the Philosophical Transactions, I found that I had been anticipated by Mr. VAN BREDA, who communicated, in 1845, a paper to the Institut on the subject; his paper appears in the Comptes rendus under an erroneous title, which accounts for its having been overlooked; he does not give thermometric measures of the heat he obtained, nor did he produce heating effects by a permanent steel magnet, or with other metals than iron. (Comptes rendus, October 27, 1845).

See also an earlier experiment by Mr. JOULE (Phil. Mag. 1843), to which he called my attention after my paper was read.

199, 203. The Experiments on the effects of Magnetism on the Matter magnetised, are collected by Mr. DE LA RIVE in his recently-published Treatise on Electricity, vol. i.

206. DAVY, Electricity defined as Chemical affinity acting on Masses (Phil. Trans. 1826, p. 389).
VOLTA, Electricity excited by the mere Contact of conducting Substances (Phil. Trans. 1800, p. 403).

207. GROVE, Gold-Leaf Experiment (Comptes rendus, Paris, 1839, p. 567).

208. GROVE, Voltaic Action of Sulphur, Phosphorus, and Hydrocarbons (Phil. Trans. 1845, p. 351).

PAGE
208. GROVE, New Voltaic Combination (Phil. Mag. vol. xiv. p. 388; vol. xv. p. 287).

GROVE, Electricity of Blowpipe Flame (Proceedings of the Royal Institution, February 1854), Phil. Mag.

211. DALTON, New System of Chemistry, London, 1810.

212. I have here and elsewhere used whole numbers, as sufficiently approximate for the argument, but without intending to express any opinion as to the law of PROUT.

213. FARADAY, Definite Electrolysis (Phil. Trans. 1834, p. 77).

216. WOOD, Heat disengaged in Chemical Combinations (Phil. Mag. 1852).

218. ANDREWS, Phil. Trans. 1844, p. 21.
HESS, Poggendorff's Annalen, Bd. lii. p. 107.

220. FAVRE, Ann. de Ch. et de Phys. vols. 39, 40; Comptes rendus, Paris, vol. 45, p. 56, and vol. 46, p. 337.

228. CATALYSIS by Platinum (DOBEREIMER, Ann. de Ch. et de Phys. tom. xxiv. p. 93; DULONG and THENARD, Ann. de Ch. et de Phys. tom. xxiii. p. 440).

229. GROVE, Gas Voltaic Battery (Phil. Mag. February 1839, and December 1842; Phil. Trans. 1843, p. 91).

231. MOSOTTI, Forces which regulate the Internal Constitution of Bodies (Taylor's Scientific Memoirs, vol. i. p. 448).

232. PLÜCKER, Repulsion of the Optic Axes of Crystals by the Poles of a Magnet (Taylor's Scientific Memoirs, vol. v. p. 353).
Magnetic Action of Cyanite (Lit. Gaz. 1849, p. 431).

233. MATTEUCCI, Correlation of Electric Current and Nervous Force (Phil. Trans. 1850, p. 287).

234. CARPENTER, On the Mutual Relations of the Vital and Physical Forces (Phil. Trans. 1850, p. 751).

236. On Effort. See BROWN, Cause and Effect; HERSCHEL's Discourse; and QUARTERLY REVIEW, June 1841.

237. HELMHOLTZ, Muller's Archives, 1845; MATTEUCCI, Comptes

rendus, Paris, 1856; BECLARD, Archives de Médecine, 1861.

256. DULONG and PETIT, Relation between Specific Heat and Chemical Equivalents (Ann. de Ch. et de Phys. tom. x. p. 395).

257. NEUMANN, Poggendorff's Annalen, Bd. xxiii. p. 1.
AVOGADRO, Ann. de Ch. et de Phys. tom. lv. p. 80.

NOTES AND REFERENCES
TO
CONTINUITY.

281. HERSCHEL, Sir J., Astronomical Observations at the Cape of Good Hope, 1847.
ROSSE, Earl of, Observations on the Nebulæ (Phil. Trans. 1850, p. 499).

282. OLMSTED, Silliman's Journal, July 1834, p. 138.
The first suggestion of a perspective vanishing point for meteors seems to be due to Professor Thomson, of Nashville.
HERSCHEL, ALEXANDER, Reports of the Meteor Committee of the British Association.
SORBY, Idem, and Proc. R. S. June 16, 1864.
BRAYLEY, Idem, 1865, p. 140, and Proc. R. S. March 23, 1865.

283. LEVERRIER, Intra Mercurial Planets (Comptes rendus, Paris, 1861, p. 1109).

284. DAUBRÉE, Comptes rendus, Paris, 1866 (Bulletin de la Société Géologique de France, Mars 1866.)

287. PLÜCKER, Variation of Spectrum Lines with Temperature (Phil. Trans. 1865, p. 6.)

PAGE
288. HUGGINS and MILLER, Spectra of Fixed Stars (Phil. Trans. 1864, p. 413).
Spectrum of Temporary Star (Proc. R. S., No. 84, 1866).
HUGGINS, Spectrum of Comet, 1866 (Proc. R. S., No. 80, 1866).

291. CHACORNAC, On the Moon (Comptes rendus, Paris, June 1866, p. 1406, &c.).

297. RUMFORD, Heat of Friction (Phil. Trans. 1798, p. 80).
DAVY, Idem, West of England Contributions, p. 18.

298. SABINE, Magnetism and Solar Spots (Proc. R. S. 1865, p. 491).

299. AIRY, On Solar Magnetism (Phil. Trans. 1863, pp. 313-646).
CHAMBERS, Idem, Phil. Trans. 1863, pp. 514-516.
MAYER, Friction of Tidal Wave (See his Papers collected and translated by Youmans, New York, 1865).

300. DELAUNAY, Acceleration of Moon's Motion (Comptes rendus, Paris, Dec. 1865, Jan. 1866).
AIRY, Idem, Notices R. Ast. Soc., April 13, 1866.
CARRINGTON, Observations on Spots on the Sun, 1863.
DE LA RUE, STEWART, and LOEWY, Idem 1865.

301. FAYE, On the Dynamic Theory of Solar Heat (Comptes rendus, Paris, Oct. 1862, p. 564).
Constitution of Sun, Motion of Sun Spots, &c. (Comptes rendus, Paris, Jan. 1866, &c).

302. STRUVÉ, Etudes d'Astronomie Stellaire, 1847.
The passage in the text is so brief as to be obscure (See the idea elaborated, Corr. Phys. Forces, p. 187).

303. See Corr. Phys. Forces, p. 84.

304. BERTHELOT, Formate of Potash (Institut. 1864, p. 332).
TYNDALL, On Radiant Heat (Phil. Mag., Nov. 1864; Phil. Trans. 1866).

305. GRAHAM, Dialysis of Air (Phil. Trans. 1866, p. 399).

306. WILDE, Increase of Magneto-Electric Force (Proc. R. S., April 1866, p. 107).

PAGE
306. HOLTZ, New Electrical Machine (Pogg. Annalen, 1865, pt. 1, p. 157).
308. CARPENTER, Food and Force; Physiology, Treatise on.
BENCE JONES, Idem, Proc. R. I. March 23, 1866.
PLAYFAIR, Idem, Proc. R. I. April 28, 1865.
E. SMITH, Idem, Phil. Trans. 1861, p. 747.
FRANKLAND, Idem, Proc. R. I. 1866.
309. TRAUBE, Virchow's Archiv, vol. xxiii. p. 196, &c.
FICK and WISLICENUS, Idem, Phil. Mag. June 1866, Supplement.
311. LAVOISIER, Œuvres, vol. ii. p. 640.
312. ANSTED, Intellectual Observer, Aug. 1864.
313. RAMSAY, Addresses to the Geological Society, 1863 and 1864.
314. HERSCHEL, Sir J., Geological Effects of Variation in Earth's Orbit (Trans. Geol. Soc., 2nd series, vol. iii. p. 295; Outlines of Astronomy, 1864, pp. 233-235).
CROLL, Idem, Phil. Mag., Aug. 1864, and Apr. 1866.
320. PASTEUR and POUCHET, On Spontaneous Generation (Comptes rendus, Paris, 1863 to 1865 inclusive).
CHILD, Proc. R. S. 1865, p. 178.
326. CARPENTER, On Foraminifera (Phil. Trans. 1856, p. 227; 1860, p. 584).
328. H. BATES, Butterflies of South America (Trans. Linn. Soc. vol. xxiii. p. 495).
330. WALLACE, Butterflies of Malay, &c. (Trans. Linn. Soc., vol. xxv. p. 1).
WALSH, Proc. Entom. Soc. Philadelphia, 1864, p. 403.
331. FRITZ MÜLLER, Für Darwin, Leipsig, 1864 (Annals and Magazine of Natural History, 1865).
LUBBOCK, Diving Hymenoptera (Trans. Linn Soc. vol. xxiv. p. 135).
LOGAN, Eozoon, Communication to the British Association at Bath, 1864.

NOTES AND REFERENCES.

PAGE

A. DECANDOLLE, Variability in Oaks, &c. (Bibl. Univ. de Genève, Nov. 1862).

331. HOOKER, On Oaks (Trans. Linn. Soc. vol. xxiii. p. 381).
332. On Arctic Flora (Trans. Linn. Soc. vol. xxiii. p. 251).
342. DARWIN, Origin of Species through natural Selection, 1866, in which see also Dr. McDonnell's results.
HUXLEY, Address to the Geological Society, Feb. 21, 1862.
LYELL, Antiquity of Man, 1863.

www.ingramcontent.com/pod-product-compliance
Lightning Source LLC
Chambersburg PA
CBHW030406230426
43664CB00007BB/767